THOMAS CRUMP used to teach Anthropology at Amsterdam University. He is the author of *A Brief History of Science*, *The Anthropology of Numbers* and *Asia Pacific: A History of Empire and Conflict*.

Titles available in the *Brief History* series

A Brief History of 1917: Russia's Year of Revolution
Roy Bainton
A Brief History of the Birth of the Nazis
Nigel Jones
A Brief History of British Sea Power
David Howarth
A Brief History of the Circumnavigators
Derek Wilson
A Brief History of the Cold War
John Hughes-Wilson
A Brief History of the Crimean War
Alex Troubetzkoy
A Brief History of the Crusades
Geoffrey Hindley
A Brief History of the Druids
Peter Berresford Ellis
A Brief History of the Dynasties of China
Bamber Gascoigne
A Brief History of the End of the World
Simon Pearson
A Brief History of the Future
Oona Strathern
A Brief History of Globalization
Alex MacGillivray
A Brief History of the Great Moghuls
Bamber Gascoigne
A Brief History of the Hundred Years War
Desmond Seward
A Brief History of the Middle East
Christopher Catherwood
A Brief History of Misogyny
Jack Holland
A Brief History of Medicine
Paul Strathern
A Brief History of Mutiny
Richard Woodman
A Brief History of Painting
Roy Bolton
A Brief History of Science
Thomas Crump
A Brief History of Secret Societies
David V. Barrett
A Brief History of Stonehenge
Aubrey Burl
A Brief History of the Vikings
Jonathan Clements

A BRIEF HISTORY OF

THE AGE OF STEAM

THE POWER THAT DROVE THE INDUSTRIAL REVOLUTION

THOMAS CRUMP

ROBINSON
London

Constable & Robinson Ltd
3 The Lanchesters
162 Fulham Palace Road
London W6 9ER
www.constablerobinson.com

First published in the UK by Robinson,
an imprint of Constable and Robinson Ltd 2007

A copy of the British Library Cataloguing in
Publication data is available from the British Library

ISBN: 978-1-84529-553-0

Printed and bound in the EU

3 5 7 9 10 8 6 4 2

CONTENTS

LIST OF MAPS

PREFACE

In 1896 my father was born into the last generation that truly belonged to the age of steam. As a little boy in Hampstead – as recounted in the children's stories written by his mother – he always took with him to bed a toy locomotive known as Yellowy. Local traffic was horse-drawn but there was also Hampstead Heath Station at the bottom of East Heath Road, on the North London line, whose construction is a constant leitmotiv of Dickens's *Dombey and Son*. Steam trains, on a busy suburban service, were the order of the day, so passengers must have been greatly relieved when their train finally came out of the long tunnel from Finchley Road on the other side of Hampstead. All this was part of my father's early years, but when he was eleven the situation changed radically with the opening of the Charing Cross, Euston and Hampstead Railway, deep underground, with electric locomotives and a new station almost at the highest point of

Hampstead. At the station entrance a crossing-sweeper was always ready to earn the odd penny by making a clean passage to the other side of Hampstead High Street: in spite of Karl Benz's invention of a motor car more than twenty years earlier, few vehicles powered by the internal combustion engine were to be seen in London in the 1900s. For long distances, steam was still the order of the day, both for trains and ships, as my father's family found when, in 1909, they moved house to Rome for a year in connection with my grandfather's work as a keeper – in 2007 aka curator – at the Public Record Office. This employment also led to one of the more remarkable train journeys of the age of steam. With the outbreak of World War I in 1914, the most valuable records were moved out of London for safekeeping in Bodmin Castle in Cornwall. My grandfather, with a remarkably free hand, was entrusted with the task of bringing the *Domesday Book* safely to this new home. His chosen strategy was to have it wrapped up in a brown paper parcel, which he himself, together with a picnic hamper, would take to Bodmin by train, with a reserved first-class seat. The incident is recorded in the official *History of the Public Record Office*, but I sometimes wonder what would have happened if my grandfather had said to a fellow passenger, 'You know what's in the parcel on the luggage rack? The *Domesday Book*'. By this time Great Western trains had had corridors for some twenty years, so I also wonder how concerned my grandfather was when he left the compartment to answer the odd call of nature. After all, the journey lasted almost the whole day. The last stage, from the station to the castle at Bodmin, was almost certainly by some horse-drawn carriage, while from his office in Chancery Lane to Paddington Station my grandfather doubtless relied on one of the hundreds of taxi cabs plying their trade on the streets of London.

In the years between the two world wars steam traction reached its highest point on the London and North-Eastern Railway (LNER), whose chief locomotive engineer, Nigel Gresley, designed the fastest steam locomotives ever made, with the Mallard, on 3 July 1938 – with a train of six carriages – attaining the all-time record speed of 126 mph. During the 1930s I fell asleep listening to the rumble of the LNER expresses on the East Coast main line as they got up speed on their way north through Wood Green, a mile or two downhill from our family home in Highgate. The LNER was always our favourite railway, and somehow, after the end of World War II, when it was destined to lose its identity as a result of the railway nationalization plans of the British Labour Government elected in 1945, the directors asked my father to write its history during the war years. To my father this was a gift from heaven: along hundreds of miles of railway line every door was open, from London to the north of Scotland, and all this came with a free first-class travel pass. The greater part of the research – consisting of talking to dozens of railwaymen, footplate crew, signalmen, shunters, guards and station staff, about their life in wartime – was carried out in the summer of 1946, during the school holidays. This was part of my father's strategy, because I was recruited as his assistant – also with a free travel pass – so that for some weeks, shortly before my seventeenth birthday, I was able to observe the last days of a great company that was shortly to pass into history. Every day, whether in East Anglia, Yorkshire, Tyneside or Scotland, I would have my own itinerary, meeting up with my father in the evening for a night at an LNER railway hotel, with dinner – at the height of Britain's post-war austerity – hosted by the LNER's local managers. Every day, one interview after another, brought me face to face with men who had made their lives in the age of steam.

My father's book, *By Rail to Victory*, was published by the LNER in 1947. It was one of the company's last projects, and it was one which gave it some pride. Until writing my own book on the age of steam, some sixty years later, my father's book was my only professional – if that is the right word – venture into the realm of steam transport. His own interest in railways never flagged, and when he died, quite unexpectedly, in January 1964, he was busy writing a book about what the future held in store for British railways after the age of steam. This project, sadly never completed, assured him a busy and very happy time during the last year of his life. His example, and the knowledge and experience acquired by working with him more than fifty years ago, are for me an essential part of the background to my book. For this reason I gladly dedicate it to the memory of my father, Norman Crump. I like to think that he would have enjoyed it.

Thomas Crump,
Amsterdam

A NOTE ON MEASUREMENTS

Inch	2.56 centimetres
Foot	30.5 centimetres
Mile	1.6093 kilometres
4'8½"	1.447 metres (railways standard gauge)
Ounce	28.25 grams
Pound	0.4536 kilograms
Ton	1016 kilograms
Bushel	36.4 litres
Metre	3.3 feet
Kilometre	3,280 feet
Kilogram	2.2046 pounds

LIST OF ABBREVIATIONS

B&ORR	Baltimore & Ohio Railroad
BCC	British and Chinese Corporation
CCR	Chinese Central Railways
CER	Chinese Eastern Railway
CI	Compression-ignition
CSS	Confederate States Ship
DAB	Deutsch-Asiatische Bank
EIR	East Indian Railway
GIPR	Great Indian Peninsular Railway
GWR	Great Western Railway
HMS	His/Her Majesty's Ship
HSBC	Hong Kong and Shanghai Banking Corporation
IC	Internal combustion
JNR	Japanese National Railways
LNER	London and North-Eastern Railway
PLM	Paris-Lyon-Marseille

P&O	Peninsular and Oriental Steam Navigation Company
RMS	Royal Mail Ship
RN	Royal Navy
SEG	Schantung-Eisenbahn-Gesellschaft
SS	Steamship
UNESCO	United Nations Educational Social and Cultural Organization
US	United States
USS	United States Ship

1

INTRODUCTION: THE TRANSPORT PARADIGM

We do not have to go that far back in time from our breathless modern world to discover another world of small self-contained agricultural communities, where life depended upon maintaining a subsistence economy capable of producing a small surplus, which, on one side could be traded for goods not locally available, while on the other it was due as tribute to a small governing class whose power and wealth depended upon maintaining the system of central government.

From a very early stage money, whose production – generally in the form of metal coins – was subject to tight government control, supported this form of political economy. Where mechanical power was needed beyond that which could be supplied by human labour, the resources of nature – whether in the form of animals with greater strength and endurance than any team of men, the downstream flow of river water or simply the winds that blew almost everywhere. The balance between these various sources of power depended on local circumstances, one of which was the demand for transport – most often met by beasts of burden. Here the wheel, however ancient its invention, was decidedly marginal: the most common use for wheels was in the driving systems of wind and watermills, where they were out of sight. The wheel would only come into its own after steam, as a source of power, was developed in the course of the eighteenth century. Why this must be so is a constant leitmotiv of this book.

The model of society as it was before the age of steam, with a time-warp measured in hundreds, if not thousands of years, is extremely general. It still applies to agricultural communities not only throughout the developing world, but also – in historical times so recent that they are still within human memory – for many such communities in Europe. If the advance of civilization – a historical process with any number of set-backs – went hand in hand with the growth of cities, this inevitably required agricultural populations to consume an ever-declining fraction of the fruits of their own labour. The balance of their production was then destined for city-dwellers, among whom could be numbered, as time went on, more and more of their own lords and masters.

For hundreds of years the process could be observed in the city states of the Italian renaissance, where, beginning in the twelfth century, the agricultural economy came to depend upon the mixed cultivation – the so-called '*cultura mista*' – of grapes, wheat and olives, by share-croppers: these were known as '*mezzadri*', by reason of the fact that their landlords were entitled to half the produce of their labour. This was not necessarily a poor bargain for the *mezzadri*, for the landowner's *fattoria* was also essential for milling their wheat and pressing their grapes and olives – small-scale industrial operations that are essential stages in the production of bread, wine and oil, the traditional staples of the Mediterranean diet. This even allowed the *mezzadri* a small surplus that could be sold on the market, which was, needless to say, the destination of almost the whole of the landlord's share – with, in this case, city-dwellers as the ultimate consumers.

To see what all this involved in the way of transport it is best to zoom in on one particular city, and the republic which it governed. For this purpose I have chosen Siena, a city where I once lived in the heart of Tuscany. Except for a number of market gardens (which are still cultivated) there was no agriculture within the city walls: the urban population had to be supplied from outside, which meant, for the most part, the *mezzadrie* were located within the republic of Siena but outside the city's walls. According to the season, and the days of the local markets, steady streams of traffic, consisting almost entirely of men and women on foot and pack animals, entered from every direction. Both people and animals would be carrying loads, generally as heavy as they were capable of carrying – for time spent on travel was lost to other productive activity. Among the travellers there would be occasional riders on horseback, who would also be conspicuous for the wealth of their attire;[1] the whole scene is portrayed in Ambrogio Lorenzetti's vast painting entitled *The Effects of Good Government* in Siena's Palazzo Pubblico, in which such details are to be noted as a flock of sheep brought within the city walls, and a single woman leading a pig on the road outside. Conspicuous by their absence are wheeled vehicles in any form.

The horse and cart, although not unknown, were not often to be seen in renaissance Tuscany, and the same was true of most of the rest of Europe. At a time when the old roads of imperial Rome had often become, by reason of centuries of neglect, impassable for such traffic,[2] vehicles on wheels were only useful in special situations, mostly involving transport over short distances, within a single domain, where their operation justified the costs of maintaining passable roads. On an estate comprising several *mezzadrie*, newly harvested grapes from the more distant vineyards might be brought to the *fattoria* by horse and cart where the terrain lent itself to the construction of adequate roads.[3] This was, however, extremely problematic, given the hazards of extreme weather and the tendency of heavily loaded axles to churn up the surface. By the sixteenth century the introduction of lighter wheels with spokes,[4] free to revolve on their own bearings, lessened this hazard, but it was only towards the end of the eighteenth century that improved all-weather roads reduced it significantly.

Whatever the use of wheeled transport in agriculture, it was much better suited to enterprises such as mining and quarrying, where heavy loads, at intervals of time measured in minutes, had to be transported over short distances, over terrain with a single proprietor – that is, the owner of the mine or quarry. Such use is recorded in England as far back as the fifteenth century, when wagons – to cite one instance – are recorded as carrying limestone some 13 miles from Roche Abbey to Sheffield.[5] Although the distance is relatively long, the road had the advantage of lying entirely within the lordship of Hallamshire, so that there was no divided responsibility when it came to maintaining it, or any possibility of tolls being levied. The point is important, simply on the principle that the strength of a chain is in its weakest link. England, and much of the rest of Europe, suffered from the delegation of the responsibility for roads to the smallest units of local government – which, in early modern England meant the parish, or in Tuscany, the *comune*. The considerable variation, from one locality to another, in the economic advantage in maintaining good roads, meant that the longer the distance to be travelled the more uncertain it became that a journey could be successfully completed – particularly in winter. The problem was exacerbated by the fact that links between local networks were seldom a first concern of those responsible for them – as can still be observed in the twentieth century by driving by car across state lines in the US. Almost everywhere in early modern Europe, when it came to shipment over long distances, the main local interest was in finding the best route to a navigable waterway – so that the problem of long-distance land transport was simply bypassed. An example is to be found in the Carrara marble, which, having been brought down to the nearby coast by river, was then transhipped to reach Rome by the Tiber – where, among many others, Michelangelo used it for his well-known sculptures. Substituting the Thames for the Tiber, much the same story can be told, in England, of the limestone from Roche Abbey in Yorkshire which was used in the construction of Windsor Castle.[6]

The advantages in unit costs offered by water transport were greatest when the goods to be transported had a low price-weight ratio: this explains the strong preference for such transport for building materials, such as stone and brick, and, as once abundant local supplies were exhausted, wood. Above all, the use of coal, which by its nature, is consumed in large quantities, would never have been economic, if it could not be transported by water. On the other hand commodities with a high-price weight ratio put a premium on transport overland, particularly when local demand was very widespread: in England this was pre-eminently true of wool, finished textiles and small mass-produced industrial products such as nails, which, even in the eighteenth century were carried everywhere by packhorse.[7] In this case the relatively small maximum load of a single horse meant that trains could divide continually, so that at the end of the day a single carrier with only two or three horses could represent the only supplier, from his own part of the country, to a distant village market. The term is extremely general: a carrier could well be a part occupation for a farm-worker in a slack season, either carrying the load himself of using his own horses to do so.[8]

In spite of its variety the overland traffic in merchandise, as described above, must still be seen in the context of rural communities organized, economically, on the basis that consumer goods were locally produced. For the mid seventeenth century this is well illustrated by Sir Edmund Verney's household at Claydon, some 50 miles north of London – where his descendants still live. The numerous inhabitants of the estate, who were self-sufficient in beer, bread, butter, livestock, fish and game birds, also sawed their own planks and forged their own ironwork, while the women spun wool and flax, made clothes, cooked, distilled herbal medicines and made wine from currant, cowslip and elder.[9] Even the domestic economy was dependent upon the outside world for some raw materials, notably iron and other base metals, and particularly in much of continental Europe, wool – often an import from Britain. Much more critical, from the sixteenth century onwards, were local shortages of wood, used both as a fuel and as the one essential construction material for buildings, furniture, machinery or transport. The consequences, for transport, are presented in Chapter 3.

A century later the domestic interior, beginning with the great houses such as Claydon, was also changing, with carpets, framed portraits and marble sculpture adorning the main rooms, so that the great families, followed by the middle classes, had every interest in better roads and vehicles for the transport of goods. Those with sufficient means also indulged new tastes, whether for coffee or clothes adorned with fur: by the end of the century the fashion in beaver hats, to be noted in any number of contemporary portraits, supported the trade of the Hudson's Bay Company, founded in 1671, to exploit the remoter regions of what is now Canada – but was then known as Rupertsland. This was the start of a consumer revolution, which in the eighteenth century would extend to a much greater part of the population, showing up, ever more forcefully, the shortcomings of state-of-the-art transport overland.

So much for the carriage of goods by land, but what about people? The answer is that most of the common people never moved further than a day's travel on foot away from home – they had little to seek further afield. It has been said that those who travelled further were pilgrims, slaves, soldiers and merchants – and by the seventeenth century only the latter two categories were of any importance to overland transport. Louis XIV, King of France, built *routes royales* for rapid deployment of soldiers, but this was because, throughout his long reign, he fought hard to extend his kingdom at the cost of his neighbours – particularly in the Netherlands. In the eighteenth century, military roads were built in Scotland for defensive purposes, and here once again the threat came indirectly from France.

Lord Macauley, looking back on the Battle of Killiecrankie (in Perthshire, 1689), relates how 'experience ... taught the English Government that the weapons by which the Highlanders could most effectively be defeated were the pickaxe and the spade'.[10] Later the Duke of Atholl, with great difficulty, built a road through the pass which would just take his carriage. (The present road, A9, is a four-lane motorway.) In all these cases there would still have been little wheeled traffic, except, locally, for heavy artillery. (Even then wheeled transport can be dispensed with as the Vietcong General Giap, fighting against the French, showed in 1954 when he placed heavy artillery in the hills surrounding their stronghold of Dien Bien Phu in North Vietnam: this was sufficient to defeat the French General Navarre.[11]) The soldiers marched on foot, with only their officers privileged to ride a horse. This reflected the general order of society, so that almost everywhere, it was the men of means and power who rode – but with their retainers following on foot. Women and children, even from the wealthiest families, stayed at home, and even daughters mostly found their husbands from close by – just read Jane Austen. Carriages and coaches, carrying not only the owner but his family, only came into their own in the eighteenth century, but even then they were mostly to be found in cities. The great Scottish engineer, James Watt's, first journey from Glasgow to London, in 1767, was on horseback. For his second journey, a year later, he was able to use the recently introduced stagecoach service. At any earlier stage, both the discomfort inherent in their design and the state of the roads they had to travel discouraged the use of wheeled transport for travellers.

Merchants, whose livelihood depended upon the bottom line, adopted the most economical means of transport, given the circumstances of their commerce. Where this meant carriage overland, their wares were mainly carried by trains of pack animals, although a prosperous merchant might well ride his own horse. In some cases, such as the import trade based on York, casks of wine brought by sea were distributed from the river-head on the Ouse by horse and wagon, but these were the exception rather than the rule. Historically the transport of casks of wine was probably the earliest use – recorded in a manuscript of 1275 – for the two-wheeled French *charette*:[12] Peter Bruegel the elder's *Census at Bethlehem*[13] shows two such *charettes* as they were some 300 years later – apparently little changed in design. In Eastern Europe horse-drawn sleds, used for transport in winter, were often more economical than any alternative wheeled transport in the summer months. Finally, even as late as the nineteenth century, driving cattle on the hoof, sometimes over distances of hundreds of miles, was a key element in supplying food to the Netherlands, Germany and Italy from Scandinavia and Eastern Europe,[14] and to London and other great market centres from many parts of Britain.[15]

Finally the most important of the every-day travellers along the roads of both Britain and the Continent were those who carried mail, according to a system that was organized on the basis of fixed stages – of such a length that each one of them could be traversed in one day by a rider on horseback. Any road, therefore, that comprised successive stages, was bound to be important: British examples are the Great North Road and the Dover Road. According to the statute by which Henry VIII established the Royal Mail in 1516, 'posts', such as carriers of mail were originally known, were contractors to the state, enjoying priority rights to essential services – at every post-town – such as a horse for every stage, at a rate fixed by the government, together with board and lodging. The French royal monopoly, created in 1464, was earlier, and depended, also, on postal carriages, 'large enough so that occasional passengers of social or political prominence might be transported between French cities'.[16] The service was minimal, with arrangements being made ad hoc: significant improvement only came with the introduction of the *diligence* in 1690. Although, until the mid-sixteenth century, the 'posts' were restricted to handling official government correspondence – and at rates which were quite unremunerative – long before the end of the century they were allowed to set up innkeepers, enjoying the title of 'postmaster', while entrusting the actual carriage of mail to servants, who at the same time served the general public – a practice officially sanctioned by Charles I in 1635. In this way the post-boy, riding his horse at a gallop and blowing his horn so that everyone knew he was coming, became a familiar figure along the post-roads. In Britain the carriage of mail was finally organized as a public service with the establishment of the General Post Office by Charles II in 1660, a reform followed in due course by other European states, where the post horn became the accepted symbol of the whole service – as can still be seen on many post office letter-boxes. To this day the carriage of mail, whether by land, sea or air, has always been both an important factor in defining the regulatory role of the state and in operating the transport infrastructure.[17] The traffic is distinctive for a number of reasons: volume, measured in terms of the number of articles carried, is very high, but each one of these is of extremely light weight with little essential value apart from the message it conveys. The most likely destination for a letter, once delivered and read by the addressee, is the waste-paper basket. On the other hand, the value of the communication between the sender and recipient can be, and often is, disproportionately high in relations to the costs of the service, which enjoys, therefore, in the jargon of modern economics, a very high consumer surplus. While this makes a monopoly – such as that enjoyed until recently by all modern states – particularly valuable, it also puts a premium on speed and reliability, with concomitant demands both upon the diligence and honesty of those employed in providing the service, and upon the standards of the means of transport.

It is now time to look more closely at transport by water, of which the decisive economic advantages have already become apparent in many different contexts. These are clearly reflected in the history of art: any gallery with a collection of old masters from Italy or the Netherlands will exhibit countless scenes containing ships of every shape and size, both at sea and crowding every harbour. The dominant role of transport by water is also reflected in the abundance and detail of written records. For certain commodities, notably grain, it was essential, as noted by the great French historian of the Mediterranean, Fernand Braudel:

> it was the inner region of the Mediterranean, with easy access to shipping routes, which could best afford the luxury of a grain trade. This in itself would suffice to explain why only those cities with direct sea links ... grew and developed.[18]

The American historian, F.C. Lane, referring to Braudel, notes, more generally, though perhaps with some exaggeration, that:

> transport by land was a hundred times more expensive than transport by water. Even as late as the American revolution it cost as much to move a ton of goods about 30 miles overland in the new nation as 3000 miles from Europe to America.[19]

Transport by water has an inherent advantage of scale. Only two factors limit the size of a vessel, and derivatively, therefore, the dimensions of the cargoes it can carry: the first is the depth of water; the second, the power needed for propulsion. As to the draught of sea-going vessels the first factor is mainly critical when it comes to finding an anchorage for transferring cargoes, convenient for access to their final destinations. Before the age of steam this meant, more often than not, transhipment to a vessel travelling inland waterways. On any coastline the ideal is a natural deep-water harbour, sheltered from the strongest winds, with a shore suitable for the construction of docks. (This was later to be important for those parts of the New World, such as Sydney and San Francisco, where economic development only started in the nineteenth century: both these magnificent harbours would have served no purpose without the coming of railways.) In practice a river estuary, which, if necessary can be deepened by dredging, is – and always was – a better alternative economically, as is illustrated by the port of Antwerp in Belgium, which for centuries has provided a key gateway to northern Europe. When it comes to the power of sea-going boats, the essential limitation is in the size of the sails – which in turn determines the dimensions of the masts and rigging that carry them. For almost the entire age of sail this factor limited the weight of ships to something over a thousand tons; this did give them, however, the advantage of a relatively shallow draught, enabling them to sail much further up rivers, to places as distant from the sea as Paris or Cologne. Sooner or later, however, cargoes would have to be transhipped to smaller boats, whose further progress upstream would be by horse-power. Even so there was still no advantage in changing to travel overland, not even onto a road suitable for horse-drawn traffic. The figures speak for themselves: where a ton was the maximum load of a wagon that could be pulled by a single horse, 30 tons was that of a barge which the same horse could tow up a river. On a canal the maximum load would increase to 50 tons.[20]

Finally, in continental Europe, wood – everywhere an essential commodity – was floated down great rivers in trains of rafts, which occasionally carried other goods, such as salt, wax and base metals. The most important traffic was on the upper Rhine and the Vistula, which, together with their tributaries, provided access to vast forests – which satisfied the demand for wood as far afield as the Netherlands. One result was that the weight of traffic downstream was three times that of the traffic upstream. In France there was similar traffic on the Seine (which satisfied Paris' demand for wood for construction) and the Rhône (which met the needs of shipbuilding in Marseilles). In Italy, wood from the Alps descended the Po and its tributaries – a traffic upon which the Swiss levied tolls and tariffs so high that they led, on one occasion, to war with Milan, whose arsenals – the largest in Europe – were gluttons for wood.[21] This form of traffic is still to be seen outside Europe: any visitor to Ottawa, looking down from the park next to the Canadian parliament buildings, will see rafts of logs slowly floating down the river and stretching as far as the eye can see.

For all its advantages, it was hardly plain sailing when it came to travel by water in pre-modern times. At sea there was always the danger of shipwreck, whether from hidden rocks and reefs off an inadequately charted coast, or from wind and storm becoming too much for a small, poorly maintained wooden ship. Distance could also be an enemy for a ship blown off course or simply becalmed, leaving supplies to run down and the crew to fall sick. Piracy was another threat to a safe and profitable voyage. Worse still, in the seventeenth century, was war at sea between nations aggressively advancing their own trade worldwide: the three wars between the English and the Dutch, in the twenty odd years following England's enforcement of the Navigation Act of 1651, led to the most hard-fought of these battles at sea, but the other great sea-going nations – Spain, Portugal and France – were at one time or another also involved in them. Only in the eighteenth century – with Britain in effective command of the seas – were these hazards brought under control, while at the same time better maps, nautical almanacs, and above all instruments – of which John Harrison's H4 chronometer is probably best known – transformed navigation. James Cook, who took the H4 on his second long voyage of exploration, related how 'our never failing guide, the Watch', performed triumphantly.[22]

Inland waterways presented quite different problems. Navigation ended where a river became too shallow or too narrow, the exact point where this happened being liable to vary with the weather: passage upstream could also be blocked by rapids and waterfalls, which could only be circumvented by constructing expensive locks and canals. Such places were often the site of watermills, with owners only too ready to levy exorbitant charges on river traffic for the facilities provided.

Flooding was another hazard, while in winter rivers could freeze. Some of these drawbacks were overcome by the construction of canals, but this was expensive, although by the beginning of the eighteenth century it was justified by the demands of new traffic – mainly in coal. In this way, both in England and continental Europe, the network of inland waterways was greatly extended to serve new industry in areas with poor access to navigable rivers. The costs, however, were very high.

Traffic on inland waterways, as much as that overland, was also subject to a multiplicity of tolls, taxes and tariffs – so much so that this was a major factor in assessing the profitability of new investment. Locally, such impositions were often a considerable brake on economic development, to the ultimate loss of those who imposed them. (In the sixteenth century Amsterdam, in extending its trade inland, benefitted considerably from the tariffs imposed by the Bishop of Utrecht on water-borne traffic in his own province.)

The fact that across the world's oceans ships were – as much for passengers as they were for goods – the only means of transport, still leaves open the question as to who needed, or was constrained to travel in the first place. Apart from ships' crews, and those accompanying cargoes, the number of those travelling by sea was notably smaller than that of those travelling for long distances – measured in days – overland. A distinction must be made here between short sea-crossings and ocean travel. Typical of the former were the links between Britain and the Continent, of which one – that between Harwich and the Hook of Holland – enjoyed a regular scheduled service from the middle of the seventeenth century. (The only interruption was during the five years (1940–5) of the German occupation of the Netherlands during the Second World War.) In the eighteenth century improved services in the short sea-crossings – combined with much better roads both in Britain and on the Continent – attracted an increasing number of wealthy travellers making the 'grand tour'.

In the case of ocean travel, two categories, emigrants and slaves, together accounted for the majority of passengers – at least in the Atlantic. Few of them expected ever to return home. In the eighteenth century soldiers crossed the oceans to garrison colonial outposts, and on occasion to fight wars – such as those against the French, in Canada and India, during the Seven Years' War (1756–63) and against the colonists in the American War of Independence (1775–83). Otherwise regular travellers were few and far between, although in the second half of the century their numbers increased significantly to include such notable American figures as Benjamin Franklin, John Adams and Thomas Jefferson.

As for inland waterways, these were not generally convenient for passengers. The one exception was the Netherlands, where the network of rivers and canals supported a considerable traffic in passenger-carrying *trekschuiten*,[23] drawn by horses on tow-paths. Otherwise travel by water was confined to ferries across rivers, where the length of the crossing was counted in minutes. Bridges – expensive to build and maintain – were few and far between.

This introductory chapter describes, in all its different modes, state-of-the-art transport before it disposed of any source of power, apart from that supplied by nature: the labour of man and beast on land and the force of the wind at sea. The history it relates ends, therefore, with the eighteenth century. This century also was the first of the steam age, but it was only at its end that the power supplied by steam was usefully applied to transport. The crying need for this result to be achieved, and the long and complicated journey to success, are the subject matter of Chapter 4. Well before the end of the eighteenth century, in much of England and continental Europe there was no room for further improvement without some technological breakthrough. The century was one of unprecedented large-scale investment in infrastructure, a process foreshadowed in the seventeenth century by such projects as the draining of the fens of East Anglia under the supervision of the Dutch engineer Vermuyden, following James I's declaration, of 1621, that he was 'unwilling to allow waterlogged lands to be waste and unprofitable'.[24] Where James I had led in the seventeenth century, Frederick II (1740–86), had followed, in the eighteenth – on a much larger scale – with his massive project to reclaim the lower Oder river, which reshaped the geography of his Prussian kingdom.[25] Then, in the second half of the century the Duke of Bridgewater, in England, knowing only too well the overwhelming advantages of transport by water for bulk cargoes, led the way in building canals for transporting coal to the new factories born of the Industrial Revolution. All this went hand in hand with vast improvements both in communication and in the provision of what today are called 'financial services'.

Finally, transport belongs to economic history. At the end of the day the success of any transport enterprise depends upon the bottom line. Because its inherent nature is to cross boundaries, any such enterprise is bound to threaten the interests of those concerned to protect their own turf – at least in their own perception. Many saw the wisdom of the old adage, 'if you can't beat them, join them'. In transport success depends not only on offering the right services, but on having the right contacts – both in business and politics. This was particularly true of the eighteenth century, when every new turnpike trust needed its own Act of Parliament and every stagecoach proprietor needed the licences appropriate to his route – all this at a time when vested interests were ready to block any new initiative. France did better, with the creation of the Corps des Ponts et Chaussées, by Louis XIV on 28 November 1713. This highly professional organization, founded for both economic and strategic reasons, built highways to connect Paris with the nation's ports and frontiers.[26] Improvement of city roads, however, was left to local government agencies. Even so, in a world where on the face of it, 'who you know' counted for more than 'what you know', the time was ripe for a technological breakthrough on a seismic scale. The tectonic plates were shifting: just where, and why, define the theme of Chapter 2.

2

TRANSPORT BY LAND AND ITS LIMITATIONS

It was noted in Chapter 1 how before the steam age the use of the wheel was never more than marginal. Indeed the vast transport network of the pre-colonial Inca Empire of Peru, which astounded the Spanish conquistadors when they first encountered it in the sixteenth century, had developed within a culture in which the wheel was completely unknown.[1] If, in the mountainous terrain of the Andes it would have been of little help – the superb paved roads were built only for men on foot and trains of llamas – wheeled transport might well have been useful in the Inca cities: Cuzco, described in a despatch to the Emperor Charles V 'as so beautiful and [with] such fine buildings that it would be remarkable even in Spain',[2] had paved streets and a vast central square surfaced in fine gravel, yet all traffic, whether of men or animals, was still on foot – as it was throughout the empire. In Europe, as shown in

Chapter 1, the wheel, more often than not, was conspicuous by its absence. Wheels were most likely to be part of some mechanism transmitting the power of a waterwheel or the sails of a windmill to the function it was designed to perform. Here the range was remarkably wide, extending from grinding corn, through driving textile machinery, to pumping water – which explains the great number of windmills in Holland: among the crowd portrayed in Avercamp's well-known *Winter Scene* in Amsterdam's Rijksmuseum, no wheels are to be seen, but the windmill in the background could not have operated without them.

Wheeled transport, where it existed in a given locality, was essentially a 'niche' service, and the property common to almost the whole range of niches – whether in mining or quarrying, agriculture or forestry – was their isolation within a small domain in which distances were short. Wheeled traffic was rare on long-distance networks, such as linked medieval Paris with Ghent, Dijon and Troyes.[3] If, from the late seventeenth century onwards, good all-weather roads and improved carriage design opened the way to long-distance travel with reasonable speed and comfort – such as was enjoyed by travellers from London to Bath – the historical process was certain, sooner or later, to be killed by its own success. There was bound to come a stage when the essential infrastructure, committed to supplying carriage horses and feeding them, would no longer cope. The stagecoach, like the French *diligence*, was a vehicle for an elite consumer class at a time when items that were once luxury articles, such as tea, were entering into mass consumption. Larger vehicles, such as the horse-drawn Paris omnibus, introduced in 1828,[4] provided for the first time in history transport for persons without reference to class, but although a similar service was soon on offer in any number of other cities – starting with London in 1829 – it was no help when it came to long-distance transport. This, however, was not where the future lay.

One must go back several hundred years to find a sequence of events that did not ultimately come to a dead end. The seeds of revolution are to be found in the mining of precious metals, such as gold, silver and mercury, and base metals, such as copper, iron, lead and tin, in many different parts of the Holy Roman Empire. This was originally due to the enterprise of German-speaking colonists, commonly known as Saxons, although they came from every part of Germany and the Netherlands. Mines, in German, are '*Bergen*' or 'mountains', the logic being that that is where their hidden wealth is to be found. Unless the lode is close enough to the surface to allow open-cast mining – generally a much cheaper operation – access requires the construction of galleries to reach the actual work-face. (Mining has a vast terminology of its own, often varying according to locality in any given language domain: lodes are the geological strata containing the ore waiting to be extracted; 'seams' are the equivalent in coal-mining.) The simplest case is that of a single gallery leading directly to the open air, in which case it is known as an 'adit'. More commonly access requires the construction of a vertical shaft, with room enough for a container – generally a basket, standard in any mining district – hoisted to the surface by rope wound on a drum. This mechanism plainly requires some sort of power, but as can be seen from any number of contemporary illustrations, this can be supplied by either men, or even better, horses. Flooding has always been one of the hazards of mining, so that the shaft containing the hoist may well be needed for bringing to the surface not only the ore mined, but also for pumping out the water that almost inevitably accumulates in mines beyond a certain depth. For practical reasons a vehicle designed to transport the ore along the galleries leading from the work-face to the bottom of the shaft could not be hoisted to the surface. At a very early stage German mines often consisted of a complex of shafts and galleries, so that loads were continuously transferred, underground, from one container to another. While galleries are ideally horizontal, an incline, if not too steep, is acceptable: this is often expedient when it comes to following a lode – in which case a gallery may be known as a 'drift'. A shaft, however, must be truly vertical, so as to allow a hoist to fulfil its task of carrying the ore mined to the surface. Shafts also contained ladders for access by the actual miners. Galleries from the bottom of the shaft lead to the work-face.

A mine, to be economic, had to remove the ore from underground at a rate that kept pace with that of the miners hacking away at the work-face. The need was for a container, which, after being loaded at the work-face, could be moved, expeditiously, along the galleries. If the distances – measured at most in hundreds of yards – were short, the volume of traffic was considerable. These two factors combined to make it economic to use vehicles running on a dedicated track: the answer, some time in the fifteenth century, if not earlier, was to lay a flat wooden surface along the floor of a gallery so as to create a 'wagon-way'. (From the beginning of the eighteenth century this expedient was also adopted in Eastern Europe (including Russia) to build roads on the surface.[5])

From a very early stage, the wagon, with four wheels, was known as a '*hund*' – German for 'dog' – of which the earliest record dates from 1427.[6] The actual container was little more than an open box supported by the axles, large enough to carry a useful load of ore but small enough to fit into the galleries and be managed by a single workman. The problem of guiding the *hund* along the wooden track was first solved by using a '*leitnagel*', literally a 'lead nail'.[7] The track then had to be laid as planks in the direction of movement, supported on cross-ties cut to the width of the gallery – everything being made of wood. In the middle of the track, the *leitnagel*, projecting down from the *hund*, fitted a gap between the planks. (From the end of the nineteenth century the same principle was applied with electric trams, taking their current via a 'plough' in continuous contact with a rail embedded several inches beneath the road surface.) The *leitnagel* meant, at least in principle, that the *hund* required no steering, but only traction – which if not supplied by gravity required no more than the muscle-power of one man. The *hund*, as so conceived, enormously improved the efficiency of a mine, particularly when its dedicated track followed a drift.

At much the same time as the *hund* was introduced in mining, military engineers had developed vehicles, to be used in building fortifications, whose wheels were guided by the track in which they ran. A drawing,[8] dated about 1430, may well be the earliest record of a true railway, in which the wheels of the wagons are guided by parallel wooden rails. The basic engineering requirement was to dig ditches and use the earth excavated to build ramparts. Because the railway, *a fortiori*, was laid up the slope of the ramparts, it was more nearly vertical than horizontal: traction, therefore, depended on a system of hoists operated by cantilevers. On the other hand the distances to be covered were no more than a few feet.

In the sixteenth century the principle applied by the military engineers was adopted in mining; then, with the introduction of flanged wheels, the rails not only guided the *hund*, but supported it. This was a key development in the history of transport, upon which there has been, to this day, no significant advance in design. Laying the tracks, however, required much higher standards of engineering.

Given the undoubted advantages of underground rail transport by *hund* in mining for metal in early modern Europe, one is left asking whether, and to what extent, the system extended further afield. For a number of reasons this was a comparatively slow process. First, the galleries in a mine define a three-dimensional network, which, if not built in rock, is shored up to protect the works from hazards such as the roofs of the galleries collapsing. Although excess water is a hazard, in any viable mine it is one that is counter-acted by appropriate engineering works – which over the course of time became steadily more efficient. The general result is that gallery floors are a hard surface well able to support a railway: what is more, being underground they are well protected from wind, rain and frost. And as already noted, traffic is regular and uniform, and distances are short.

During the sixteenth century it became clear that English forests were being cleared at a rate, which was unsustainable – as is the case today in much of the tropical world.[9] For many there was an economic advantage from new land becoming available for agriculture, but that did nothing to resolve the essential crisis. The obvious solution was to find a substitute for wood both as fuel and as a construction material. The answer was to be found first in coal, which during the reign of Queen Elizabeth I (1558–1603) became England's main fuel, and then, a century or two later, in iron, the material, par excellence, for constructing the machinery demanded by the Industrial Revolution. By this time brick had also largely taken over from wood in the construction of houses, factories and public works.

When it came to transport, packhorses, although used for the purpose, were a slow and inefficient means of delivering heavy, low-cost, bulk cargoes such as coal, bricks and iron: ships, sailing both the high seas and navigable rivers, were a far better alternative. When it came to coal this made all the difference: the main demand was in London, with its miles of docks along the Thames, while the main supply came from mines along the River Tyne, more than 200 miles to the north in the counties of Durham and Northumberland. In London, a city with few hills and paved streets, horse-drawn wagons could transport coal to its end-users. Along the Tyne, where the mines were dug into the sides of steep hills, this was not a viable means for bringing coal down to the quayside.

The problem presented by the mines was the short-haul overland transport of low-cost bulk goods. However the coal reached the Tyne, economic factors dictated that the journey be as short and simple as possible. By the sixteenth century there had already been a revolution in short-distance transport as it related to mining, but it had nothing to do with coal, nor did it take place in England.

All these advantages are lost, in whole or in part, when it comes to surface transport. Possible routes for road or railways are constrained by local topography – which in mining regions tends to be a hindrance rather than a help – and distances to ultimate consumers are much longer. Other things being equal, these factors combine to make it economic to refine ore extracted from an underground mine as close as possible to the point where it reaches the surface: copper, for example, has less than a quarter of the weight of the ore from which it is extracted, and for precious metals the fraction is much smaller. The crunch comes when essential factors of production, such as water and fuel – generally charcoal – are not locally available, but this was hardly the case in the mining areas of continental Europe in early modern times. There, by reason of the local topography, mining was a much more profitable use of land than agriculture, so that local forests were left more or less intact (as, comparatively speaking, they still are today) and mountain streams provided abundant water. What is more, mining at this stage had little to gain from economies of scale, so that the production of refined metals, at or close to the surface works of the mine itself, was an integrated process.

Above ground, therefore, rail haulage tended to lose its decisive economic advantages. An illustration from 1544,[10] for example, shows a *hund* on rails coming out from underground via an adit, with in the background two much larger four-wheeled wagons – one drawn by horses and the other by oxen – carrying away the finished product of the mine. Their probable destination was a staithe – the term for a 'wharf or staging where wagons are unloaded into boats' – on the nearest navigable river. In different circumstances such wagons could well have been replaced by pack-animals.

Whereas, until the end of the sixteenth century, almost all railways were to be found inside the mines of continental Europe, in the course of the seventeenth century the action would switch to England, where – on a scale much larger than anything seen on the continent – new railways were built to transport coal from the pitheads down to the most convenient navigable waterway.

For a long time after Britain began to mine coal underground, the means of transport from the coalface to the shaft was much more primitive than in the continental metal mines. Until well into the eighteenth century the most common practice was to load, at the coalface, a wicker 'corf' – a sort of basket standard in any coal-mining district – which would then be dragged on a sledge, with iron or wooden runners, to the mineshaft, to be hoisted to the surface. Unlike the continental *hund*, the sledge was not guided along its path, although planks were sometimes laid in order to reduce friction. Children – girls as well as boys in Scotland – pushed and pulled the sledge, working a twelve-hour day underground.

Gradually, as the century went on, 'Galloway' pit-ponies took over this work: [11] this allowed a sledge to carry up to four corves, substantially increasing productivity – although getting the ponies down into the mine was a considerable problem. (In the early days, when mines tended to be smaller, a special sloping adit was constructed for the ponies to enter from outside. Later, when mines became larger, they were let down the shafts in slings.[12]) Finally, vehicles with wheels, known as 'rollies' (which were always horse-drawn) or 'trams' according to their design, began to replace sledges, but rails only came into use at a very late stage. At any given time sledges, trams and rollies could all be used in one and the same mine. ('Sledges', 'trams' and 'rollies' were the terms used on Tyneside: other terms, such as 'skip' and 'dan' were used for mines along the Severn, with others used elsewhere.) The choice depended on the dimensions of the galleries along which coal had to be transported from the coalface to the mineshaft, so that a loaded corf could start its journey on a sledge, then transfer to a tram, to end up with a rolley bringing it to the bottom of the shaft, where it would be hoisted to the surface. Transport planning underground was almost invariably ad hoc, which no doubt explains why the means adopted remained primitive for so long. However it was organized, it still depended upon child labour, which continued in Scotland until the mid nineteenth century.[13]

Transport underground is not the only way in which mining coal differs from the extraction of metal ores. There are also very substantial differences on the surface. First, where coal, once above ground, may have to be cleaned and sorted, no further refinement is needed. The product is ready for use: the problem is simply to get it to the users – which is where railways come in. Second, because coal – unlike metal – is consumed in the fire, which it fuels, production must be on a much larger scale. The result for any economy – such as that of England from the sixteenth century onwards – which switches from wood to coal as its main fuel, is that mining becomes an industry requiring unprecedented levels of both labour and capital. These are needed not only for the actual mines, but even more for the transport infrastructure that will support them.

The development of coal-mining depended on exploiting not only the richest seams, but also on giving priority to those close to navigable waterways. Once coal was loaded onto a boat transport costs were no more than a fraction of those for any form of surface transport. By the end of Elizabeth's reign in 1603 it was clear that this favoured mines close to England's great rivers: given the actual location of the richest seams in the north midlands, the Welsh marches and the north-east, this meant, above all, the Trent, the Severn and the Tyne.

In 1601, Huntington Beaumont, the younger son in a family that already mined coal in Leicestershire, moved north to see whether he could profitably exploit the richer seams to be found in three parishes immediately west of Nottingham. There, in 1603, his eye lighted upon the parish of Strelley, where he obtained, from Sir Philip Strelley, a lease to work coal for fifteen years. The pithead was 2 miles from Wollerton Lane – the distribution point for Nottingham and other towns along the River Trent – and with a way leave from the owner of the land between them, Beaumont, applying the key principle 'before you buy a mine, buy a road',[14] had, by October 1604, linked them 'along the passage now laide with Railes, and with suche or the lyke Carriages as are now in use for that purpose'.

One cannot be quite certain that this was England's first ever railway, because at much the same time similar developments were taking place in Shropshire, along the Severn, but it was Beaumont's success with his 'wagonway' that pointed to the future of this means of transport. It also provided the first of countless occasions, in every part of the world, for challenging it. Even before Beaumont's fifteen-year lease expired, Francis Strelley had questioned it in the Chancery Court, claiming that

> the said Huntington Beaumont hath used new and extraordinary invencions and practises for the speedy and easy conveyance away of the said coales, and especially by breaking the soyle for layinge of rayles to carry the same upon with great ease and expedicion ... and by drawinge of certain carrylaggs laden with coales uppon the same rayles.[15]

Although the 'invencion' was not so new and extraordinary – having been applied in Germany for some 200 years – it was to transform the mining infrastructure of England. Beaumont, impatient with the likes of Francis Strelley, went up north to open new mines in Northumberland, where he built three wagonways down to the River Blyth close to where it reaches the North Sea. Although his initiative and enterprise were appreciated, he lost all his money in the end, which could have been as much as £30,000 – a vast sum for the time.[16] This was the beginning of rail transport in the north-east of England, where, in the following two centuries it would evolve into the present state of the art, based on steam locomotion.

From a modest start along the Blyth, by the end of the seventeenth century wagonways had spread along two larger rivers further south – the Tyne and the Wear – bringing down to the staithes along their banks coal mined from the richest seams ever discovered worldwide. Other wagonways led directly to the North Sea coast. Economically the area had every possible advantage: the mines were close to water, and the sea, never far away, was on the right side of England for London – the main market for coal. The Newcastle wagonway – named after Tyneside's largest city – provided a standard design, later widely adopted elsewhere in England and Scotland.

The fact that pitheads – for topographical reasons – were invariably well above sea-level, meant that wagonways could be constructed so that their route would be either level or downhill, enabling horses to bring down to the staithe a full wagon carrying up to 5 tons of coal. (The weight of coal was actually measured in chaldrons, but far from being a standard measure the chaldron varied considerably from one part of the country to another.) This at least was the principle, but in practice local topography meant that cuttings, embankments and bridges – but, not at this stage, any tunnels – were often essential for making such a route possible. As time went on, and the seams closest to the rivers were exhausted or became unworkable for other reasons, the need for such construction added considerably to the costs of the mining operation. The situation was often exacerbated by landowners along the designated route refusing to grant the mine-owners, who were almost certainly the owners of the wagonways, the necessary way leaves: this was often a matter of hard bargaining, since the landowners – as already noted in the Strelley case – appreciated only too well how much money there was to be made from mining coal.

By the early eighteenth century the engineering required by the transport infrastructure of the mines produced works on a scale which left visitors to gaze 'open-mouthed at man's mastery over Nature in the cause of trade and industry'.[17] The most imposing of all was the Tanfield wagonway in County Durham; this included the Causey Arch, a bridge built from bricks, which still stands, and carried a railway line until 1962.

More important, historically, were the wagonways leading to Willington Quay, on the north bank of the Tyne about halfway between Newcastle and the North Sea. The first section, linking a mine known as Killingworth Moor to the river, was opened in 1764, while the final section, extending the wagonway to a new mine, Killingworth West Moor, waited until 1806. This was the mine that employed George Stephenson, the great pioneer of steam-powered locomotion, who adopted the Killingworth gauge of 4 feet 8½ inches as standard for all the new railways in which he was involved.[18]

Although more wagonways were to be constructed along the Tyne than anywhere else in Britain, the Newcastle standard was soon adopted in other important mining districts. Here, on the initiative of Sir John Lowther, a considerable local landowner, the mines of the Cumberland coalfield led the way, with the first wagonway, from the Seaton pit to Whitehaven, being open for use in 1734.[19] Wagons with cast-iron wheels, based on a recent innovation in a quarry near Bath, were introduced a year later. The Seaton wagonway was followed by two much more important links – in terms of volume of traffic – opened in 1735 to Saltom and in 1738 to the Parker Pit. Construction of the Cumberland wagonways was complicated by the fact that the staithes they led to were located on the coast of the Irish sea in the county's two main harbours, Whitehaven and Workington. This had the advantage, however, of providing immediate access to a short-sea route to Ireland, the principal market for Cumberland coal. Across the sea this led the Irish government to make a grant for the construction of a new harbour at Ballycastle, on the coast of County Antrim, for the export of coal from the small local mine. Here a wagonway was constructed not to carry coal, but stone needed for the harbour works from a nearby quarry. As things worked out coal-mining never prospered in

Ireland, leaving the market to be supplied mainly by the Cumberland mines, the last of which, at Bedlington, only closed in the 1970s.

Three other key coal-mining districts, South Yorkshire, Aire and Calder, and Fife, followed on after Cumberland in the construction of wagonways based on the Newcastle standard. Of these, Fife, although the last in point of time to develop its mines, in the end constructed the greatest length of wagonways. These, as in Cumberland, led directly to harbours on the sea – in this case on the north shore of the Firth of Forth. The result, by the end of the eighteenth century, was that transport by rail, along wooden wagonways, was known in many different parts of the British isles. The wagonway networks were, however, always local: no one ever conceived of the idea of extending them to provide for long-distance transport. Since the usefulness of wagonways depended upon the principle of 'downhill all the way', geography invariably ensured that they would always be confined to the short-distance transport of coal and minerals – with the journey always ending at, or close to, a navigable waterway.

Coal-mining along the Severn, earlier in point of time than Tyneside, was always a separate case. One main reason was its close tie not so much to local iron-mining – which had a very long history – but to the revolution in the industrial use of iron which followed from Abraham Darby's process – perfected in Coalbrookdale in 1709 – of smelting iron with coke instead of charcoal. This new link between coal and iron lay at the heart of the Industrial Revolution as it developed during the eighteenth century – so much so that a 2-mile stretch of the Ironbridge Gorge in Shropshire, where Coalbrookdale is located, was inscribed on the World Heritage List by UNESCO in 1986. The location of the Shropshire coalfields along the Severn was such that adits, to a far greater extent than was ever possible in Tyneside, provided access to the coalface – so making possible direct rail links to the river. As early as 1729 flanged iron wheels were used on such lines, probably for the first time anywhere. Then, in 1767 cast-iron rails were reported for the first time, with production on an industrial scale following at Coalbrookdale a year later. To begin with these rails were laid on top of wooden rails, thereby very substantially reducing wear and tear – which had meant that such rails had continually to be replaced. (The useful life of wooden rails was never more than six years, after which they were often only good for being burnt for charcoal.[20]) Although by the end of the century this was as important an economic factor as labour, it was only after mid-century that the scarcity of local supplies of wood began to make itself felt in Tyneside, which was soon constrained to import from as far away as Sussex. From the end of the century the high price of wood as a result of the Napoleonic wars led to the conversion to all-iron rails, and even the use of stone sleepers.[21] By this time, the vastly increased supply of iron, from works such as Coalbrookdale – together with related technological improvements – had enormously reduced the use of wood as a construction material in industry. The changeover was heralded by the completion of Telford's magnificent iron bridge over the Severn in 1780.

Indirectly Shropshire also led the way in linking coal mines to canals, rather than rivers, by rail. In 1757, the opening of the Sankey Canal linking the River Mersey to St Helens in Lancashire, led a number of local mine-owners – in a relatively new coalfield – to lay railways down to it. The first of them was laid, remarkably for the time, by a woman, Sarah Clayton, with considerable industrial interests. The idea may have come to her from visiting the quarry near Bath mentioned on page 30, but be that as it may, her initiative was soon followed by other mine-owners.

As the network of new canals in Britain developed in the second half of the eighteenth century, such rail links became ever more useful. Not only did they enable coal to be brought down to staithes on completed canals, but also building materials to be used in constructing them. Short railways were also built to get round the need to build locks. Of these the most spectacular were the three 'inclines' on the Shropshire Tub-boat Canal, finally completed in 1791. The inclines were self-balancing, so that at each end, tub-boats, with 8-ton capacity, were floated onto specially designed cradles; then as one ascended, the other descended. (As an alternative the American engineer, Robert Fulton, suggested, in 1796, fitting wheels to tub-boats, and replacing all locks with suitable inclines.[22] The same principle was adopted for the Falkirk wheel, opened in 2001 as part of the new Millennium Link of the Scottish Union Canal, originally constructed at the end of the eighteenth century. The wheel, some 30 metres in diameter, rotates on a horizontal axis and carries two cradles, positioned so that when one is at the top, the other is at the bottom. The cradles hold a depth of water regulated so that when the gates are open boats can sail in and out at both levels of the canal. Given the date of its construction the wheel is powered by electricity rather than steam.) On one of the inclines on the Shropshire canal, at Wrockwardine Wood, the power needed was supplied from the start by a steam engine; the two other inclines, originally powered by men and horses, were very soon converted to steam.

By the end of the eighteenth century, the principle of steam-powered short-distance transport – as exemplified by the inclines on the Shropshire Canal – had been extended, in a number of special cases, to transporting coal over distances of a mile or more. This was achieved by the bank winding engine, which consisted basically of a large revolving drum, on which a cable, equal in length to the wagonway, could be wound. Such engines were powered by Watt's double-action steam engines, which to the end of their days remained 'atmospheric'. This meant that they were extremely large, as can be seen, for example, in one such engine of 1833, now displayed in the Railway Museum in York. Considering that the power of the engine had to be sufficient not only to haul wagons filled with coal uphill, but also to wind a very long and heavy cable, there were obvious limitations to their usefulness. Even so they played an important part in allowing trains of wagons to transport coal, a result difficult to achieve so long as wagons were drawn by horses over tracks which were far from level – the normal situation in any mining district.

The need to wind the cable could be avoided by doubling the lines to allow for counter-balancing – as on the canal inclines – but such a system was only viable within narrow constraints of gradient and alignment. It did, however, have the advantage that much of the power needed was supplied by heavy wagons, loaded with coal, descending under the force of gravity, while the wagons going uphill were empty. Even so a steam engine – even if of reduced power – was still necessary to drive the system.

The fact that bank engines were introduced at a relatively late date reflects the steady substitution of iron and steel for wood in the working parts – such as the beam. Miniaturization was only possible at the cost of a critical loss of power, so that even if a small-scale version of such an engine were actually mounted on the vehicle it was meant to move, the power generated would be insufficient to do so. This meant that Watt, and those working with him, never solved the problem of locomotion. On the other hand, because increases in scale were much easier to achieve, their stationary engines were able to provide steam power to Britain's new industry on an unprecedented scale. These, with their vast flywheels, were soon to be found in places where watermills could never have satisfied the local demand for power. In new textile mills, pulleys and leather driving belts connected any number of looms to a single steam engine – a form of transmission that survived well into the twentieth century.

The problem was to supply these new engines of industry with coal: this explains the network of canals – described in Chapter 6 – that developed in the late eighteenth century, largely on the initiative of the Duke of Bridgewater. Although the new canals enabled large-scale industry to be located in quite new areas – and at a level that led to the growth of many new industrial towns – there was plainly a limit to what could be achieved by relying on inland waterways for the transport of goods, whether raw materials (including fuel) or finished products. At the end of the day, Birmingham, the greatest of the new cities, was reputed to have more miles of canal than Venice.[23]

The new turnpikes of the eighteenth century were never going to provide an alternative transport infrastructure: their usefulness was mainly for carrying passengers and mail. Even with the civil engineering technology developed by Telford and Macadam, their roads would never be up to transporting the heavy loads shipped by water – if only because of the critical limits to the size of vehicles that could be pulled by horses. For goods, with a high value-to-weight ratio, such as tea, transport overland had definite advantages, like being able to reach small local markets far from navigable waterways. With both roads and vehicles much improved, surface transport had the advantage of speed, but for bulk goods such as coal and building materials this counted for little. Even today, such commodities can take weeks of travel by water to reach their final destination, and where available this is still the preferred means of transport. When, however, it came to an integrated system of transport, the industrial world of the late eighteenth century was crying out for a breakthrough.

3

THE DOMINATION OF SEA TRAFFIC

Towards the end of the eighteenth century the voyage of the First Fleet, carrying both settlers and convicts to Australia – a land then almost totally unknown – represents the apotheosis of passenger travel by sea and under sail. The fleet, consisting of the frigate *Sirius* – under command of Captain Arthur Phillip RN – a tender, three store ships and six transports, had a total tonnage of just over 3,000. Setting off from an anchorage off the Isle of Wight on 13 May 1787 and transporting some thousand people, the fleet reached its final anchorage in Australia – at Port Jackson, in what is now Sydney harbour – on 26 January 1778, having abandoned as unsatisfactory its first anchorage in Botany Bay.[1] In the course of its long voyage the First Fleet had called on only three harbours; Tenerife, Rio de Janeiro and Cape Town, calls that were essential for replenishing water. By the standards of the time the voyage, following the careful planning of Captain Phillip, was exceptionally successful: not a single ship was lost, and there were only forty-eight deaths at sea. A measure of this success is to be found in the fact that the fleet's largest vessel was lost within two years, while the best-equipped relief ship sank on the way out to Australia.

Although the First Fleet completed a remarkable voyage – and as such should have become the model for the two fleets that followed it in the final years of the eighteenth century – the actual achievement, judged by the performance of steamships a half century later, was unimpressive. Even so, it represented the best that could be achieved in the age of sail, as is made clear by the disastrous voyages of the Second and Third Fleets, which arrived in 1790 and 1791, after being sub-contracted to a London firm of Camden, Calvert and King, specializing in transporting slaves.[2] After 1793 the Napoleonic Wars inhibited the British government from organizing fleets for the transportation of convicts.

The limitations to which the First Fleet were subject were threefold. Consider first the route it followed: geographically speaking, Tenerife (in the Canary Islands) and Cape Town (at the southern end of Africa) were logical choices, but Rio meant a detour to the west of some thousands of miles. But then the north-east trade winds blew ships away from the west coasts of Africa, with the predominant ocean currents carrying them in the same direction. Rio, also, was a fine natural harbour – with all the stores necessary for replenishing a fleet – and sailing south the wind and currents became much more favourable, making for a relatively easy passage to the Cape of Good Hope and onwards across the Indian Ocean to the south coast of Australia. Secondly, there is the question of time: in the age of sail eight months was short for so long a voyage. Many took far longer, that is, if they reached their final destination. Thirdly, although the ships of the First Fleet could have been larger, their actual size was by no means sub-optimal by the standards of the day.

All these factors combined to add up to a very labour-intensive enterprise, although for the British Parliament in Westminster, shipping much of the convict population to the other side of the world solved many problems. Where the ships of the First Fleet led the way, many more followed – in spite of the disasters that overcame the Second and Third fleets – providing for the colonization of an almost empty continent.

Whatever the problems still encountered at sea, by the end of the eighteenth century conditions for sailing the world's oceans were significantly better than they had ever been before. Following the three great voyages, 1768–71, 1772–5 and 1776–80, of James Cook, there was little of the world's coastlines left to discover. At the same time boats were better built, and navigation had become much more reliable. Better charts, covering an ever greater part of the world's oceans and their coastlines, became available, and by the eighteenth century they were also beginning to record not only magnetic variations but also ocean currents.[3] In 1766, the Astronomer Royal, Nevile Maskeline, produced the first annual *Nautical Almanac and Astronomical Ephemeris*, which is still published every year.[4] More important still were new and improved instruments – of which John Harrison's H4 chronometer was the most important (as Cook testified after using it on his second voyage).[5] At the same time coastlines, particularly of Britain, were becoming much safer as new lighthouses were constructed on the model of Smeaton's Eddystone lighthouse, completed in 1759.[6]

Well before the end of the eighteenth century ships sailing the world's oceans enjoyed unprecedented security from hostile action, whether as a result of war or piracy. This radical change for the better from the situation in the seventeenth century was mainly the result of Britain's growing dominance as a maritime power – largely at the cost of the Dutch. When, in 1740, the Scottish poet, James Thomson, wrote 'Rule, Britannia, rule the waves' his exhortation might hardly have accorded with political reality, but, even so, his words accurately foreshadowed the future. Britain's position in the world's oceans was helped by two signal victories over the French – by Robert Clive at Plassey in Bengal in 1757, and James Wolfe at Quebec in Canada in 1759 – both incidents in the Seven Years' War (1756–63). On the long sea routes to the East Indies the once feared Dutch were tamed, while Spain and Portugal had gone into secular decline as significant maritime powers. If throughout the century their ships, mainly in the Atlantic, still suffered from privateers from France, Holland and the British North-American colonies – practising a form of state-licensed piracy finally abolished by the Convention of Paris of 1866 – this counted for relatively little at a time when the balance of world trade was shifting decisively in favour of Britain. The decline of privateering was important not only in terms of increased safety, but for the fact that heavy guns no longer needed to be carried by merchant ships, or the crews to man them. The result was a considerable increase in freight revenue combined with a useful reduction in labour costs.[7]

Over the long term Britannia's rule of the waves would be challenged as a result of the success of the American Revolution (1776–83) and the emergence of the United States as a significant maritime power, but until well into the nineteenth century the Royal Navy was quite capable of ensuring that it was respected – not least due to its victories in the Napoleonic Wars. The Dutch had learnt the hard way, some twenty years earlier, when they started running guns to the American rebels from the Caribbean island of St Eustatius. The result, in the years 1780–4, was the fourth Anglo-Dutch sea war, in which – in contrast to the three fought in the mid seventeenth century – the Dutch were consistently defeated.

What, then, were the shortcomings of sailing ships at the end of the eighteenth century? Clearly, they were too small and too slow to cope adequately with the expansion of trade and travel foreshadowed by the Industrial Revolution. In the words of a noted French historian, 'the wooden ship, propelled by wind ... could not exceed a certain size, number of crew, surface of sail, or speed.'[8] Moreover wood, by the end of the eighteenth century, was becoming an increasingly scarce and expensive commodity. Whereas in construction and public works it was being replaced by bricks and stone, and in transport, and above all, industry, by iron, such expedients were little use to shipbuilding in the age of sail. And, as noted in Chapter 2, the Napoleonic wars caused a critical shortage of wood.

More serious, in an age when timetables were beginning to regulate many aspects of life – particularly in commerce and industry – were the uncertainties born of utter dependence on natural forces beyond all human control. When it came to wind, currents and tides, the best that could be done was to take into account the interaction between such factors as were constant, and seasonal variations. By the end of the eighteenth century, in the case of tides, tables based on accurate local records were available. But there was little such help with winds and currents. When these were more or less constant in one direction, as in the Pacific between Acapulco in Mexico and Manila in the Philippines, year-round sailings were possible. Spanish galleons simply followed quite different routes – already established in the sixteenth century to take into account the direction of prevailing winds – according to whether they were sailing east or west across the Pacific. The west-bound galleons carried Mexican silver from Acapulco, and the east-bound galleons, Chinese goods traded at Manila – a trade that made a profit for Spain over a period not far short of 300 years: it ended only when the Spanish continental empire in the Americas began to disintegrate in the early years of the nineteenth century. The last voyage back to Acapulco was in 1811, appropriately by a galleon named *Magellan*.[9] The difference between the two routes is shown by the fact that the route of the westward voyages lay south of the Hawaiian islands, while that of the eastward voyages lay north of them; remarkably, in nearly 300 years the Spanish never discovered these islands – this was left to Captain Cook, in 1778, during the third of his great voyages.[10] As for annual variations, the monsoon seasons of the Indian Ocean determined the time of year that shipowners would plan to reach their various destinations in India and South-East Asia.[11] In such planning there was always considerable uncertainty: the exact timing of the monsoon seasons varied from one year to another, as did also the actual weather conditions brought by the monsoon.

By the end of the eighteenth century the international maritime community had long accepted all these uncertainties: there was no other choice. Closer to home, in the North Sea and the North Atlantic Ocean, weather conditions – whether storm or calm, east or west winds – were much more changeable over the short term, so that the time spent in waiting to enter or leave a harbour was measured in days rather than weeks or months, but even then, hours – in which the wind could well change direction – could be lost waiting for the right tide. If, in comparison to the months spent at sea in long ocean voyages, this mattered relatively little, it was quite different with short sea-crossings, such as between England and France or the Netherlands. Looking at J.M.W. Turner's (1818) picture of the *Dort Packet Boat from Rotterdam Becalmed*[12] one can hardly doubt the frustration of its crew and passengers as they waited to set sail – more than likely for England. Once the wind was right the crossing would take a matter of hours. Turner shows the opposite extreme in his *Snow Storm – Steam Boat off a Harbour's Mouth* (1842), which he painted after being lashed for four hours to the mast of the *Ariel* on the night she left Harwich.[13] But then the crossing could well have been considerably shorter than that of the becalmed Dort packet boat in the other direction, some twenty-five years earlier.

What is true of short sea-crossings applies equally to coastal shipping, although in this case the small number of passengers carried could mean that uncertain schedules were less critical. In England the heaviest traffic was that of colliers bringing coal from the Tyne and Wear to London – also painted by Turner. On the Continent the traffic in the Baltic and along the North Sea coasts was less uniform – with wood, grain, beer and furs, together with copper and iron from Sweden, constituting the most important freight. Wine was always important as freight moving eastwards up the French and Dutch coasts, counterbalanced by the products listed above moving in the other direction.

For all its importance for trade within Europe, coastal shipping, at the end of the eighteenth century, was relatively much more important in North America. Judged in terms of having a harbour open to sea-going ships – which defines much of the important 'tidewater' area of Virginia – every one of the thirteen former British colonies that in 1789 joined together to constitute the United States of America, had an Atlantic coastline. The great new cities such as Boston, Providence,

New Haven, New York, Philadelphia, Baltimore, Richmond, Charleston and Savannah were all linked by coastal traffic. Goods in the domestic exchange economy were transported by sea: except over short distances there was no real alternative. What is more, great navigable rivers, such as the Connecticut, the Hudson, the Delaware, the Susquehanna, the Potomac, the James and the Savannah, greatly increased the range of this traffic. At the same time, in a part of the world with a dense forest cover of natural hardwood, there was no shortage of wood for the construction of ships. Inevitably North American shipbuilding became an industry to rival that of Europe. What is more, it also built ships for purposes, such as whaling, which were less important in Europe.

Finally, the question arises as to the nature of the interface between sea-going traffic and that carried by inland waterways – which by the end of the eighteenth century included a considerable modern canal network. As a matter of broad principle, it had always paid to have sea-going ships sail as far as possible inland, reaching such unlikely places as Cambridge and Florence. Trans-shipment was expensive, but by the end of the eighteenth century, the volume of traffic and the increasing size of ships meant that the points at which it took place moved steadily downstream – a process that has continued to the present day, so that Tilbury, close to the mouth of the Thames estuary, is now the container port for London, while the old docks have been filled to create new urban areas. Each major river came to have its defined limits for sea-going ships, such as, notably, London on the Thames, Cologne on the Rhine and Lyons on the Rhône, or, on the other side of the Atlantic, Montreal on the St Lawrence, Albany on the Hudson and Philadelphia on the Delaware.

This leads to the question as to what sort of boats sailed upriver beyond these limits. Of necessity they were of shallow draught and relied on some sort of link to the banks of the inland waterways for propulsion. This explains the towpaths to be found along almost any navigable river in Europe, at least beyond the point accessible to sea-going traffic. Along such rivers, which were essentially non-tidal, horses – tied by ropes to the boats – provided the necessary power as they walked slowly up and down the towpaths. The amount of work demanded from the horses depended on the direction travelled – up or downstream – the size of the load, and the strength of the current. A viable transport network also depends upon a number of conditions, which cannot be taken for granted. Natural waterways are inherently unpredictable. In a season of heavy rainfall – which need not be local but can be far upstream – a river can easily flood its banks, leaving towpaths under several feet of water, to the point even that the dry-season channel is difficult to discern. A river in flood can also erode its banks, to the point that it follows an entirely new course once the flood subsides. (The Mississippi is the world's most notoriously variable river. Mark Twain, writing in the mid nineteenth century of the French explorer, Robert, Sieur de Lasalle, who went down the river in 1682, noted that 'nearly the whole of the one thousand three hundred miles of old Mississippi River which Lasalle floated down in his canoes, two hundred years ago, is good solid dry land now.'[14] There is no rival to it in Europe. The Rhine's flood plain is one fifteenth that of the Mississippi, and its discharge of water into the sea, one twenty-fifth.) Public works, on a massive scale, are the only possible counteraction. Although the principalities of the Po valley of northern Italy learnt this lesson in medieval times,[15] Northern Europe was slow to learn from them. Led by the Netherlands, where the whole economy depended upon inland waterways, England and Prussia, from the seventeenth century onwards, invested in flood control and land reclamation on an unprecedented scale – as already related in Chapter 1. Not only that, where rivers do not flood they can flow precipitously over rapids: control is possible by constructing weirs (often in conjunction with watermills powered by the current) but only locks – originally a fifteenth-century Italian invention, for a time wrongly attributed to Leonardo da Vinci, and in any case an expensive piece of engineering – can allow boats to continue their journey.

In the long run the greatest disadvantage of natural waterways is that they all too often provide little or no access to economically important destinations. Historically there was, even as late as the eighteenth century, an inherent paradox here: a location, without such access, would never be important economically.[16] With the construction – at the end of the century – of new industry, based on machinery made of iron powered by steam generated by burning coal, a ready supply of these two commodities became more important than access to natural waterways – even when the two, for geophysical reasons, were quite incompatible. This was behind the Duke of Bridgewater's initiative in building canals – already noted in Chapter 1 – which eventually led to the network of some 4,500 miles shown in the map on page 149. At a price, canals could also overcome the geographical shortcomings of any network of natural waterways: tunnels could take them through hills, aqueducts could take them across valleys, although routes were planned to avoid such costly expedients. There were still inherent limitations: traffic congestion was always going to be a problem, particularly at intersections and long flights of locks. Canals can also run dry after a long period of low rainfall. Every boat passing through a lock costs water which cannot always be taken for granted. Maintenance is constantly necessary, as witness the degenerate state of British waterways reached by the mid twentieth century. The final blow came with the great freeze of the winter of 1963, when for weeks boats were set fast in ice.[17]

Because the subject of this chapter is essentially the final years of the age of sail, its focus is on shipping along, and between the coasts – European and American – of the North Atlantic. As Chapter 6 will show, it is here that the maritime age of steam had its beginnings. The reasons, as almost always in transportation, were economic. Where the shortcomings of sail led to its being replaced by steam well before the middle of the nineteenth century in the North Atlantic, the reign of sail, particularly in coastal traffic, survived – even to the present day – in the seas of other parts of the world. Although by the end of the eighteenth century the character of sea-going shipping, worldwide, had achieved remarkable uniformity – largely as a result of ships designed for the Atlantic entering the two other great oceans from the sixteenth century onwards[18] – many seaboard states had characteristic local models designed for special circumstances, such as the change in the monsoon seasons in the Indian Ocean. These, in particular, survived the coming of the age of steam, so that even today one need not look far in the Indian Ocean to find classic Arabic dhows, or in the China seas, to find junks, both of which still have considerable cost advantages for small-scale entrepreneurs in a world where labour is still cheap – it is not for nothing that modern merchant shipping looks for its crews in such countries as the Philippines and Bangladesh. The same is true of the great rivers, such as the Ganges, the Irrawaddy and the Yangtze, where even passenger ferries – crossing from one side to another – if not under sail, may still be propelled by oars. The case is quite different with the multifarious short sea-crossings, so characteristic of the island nations of Indonesia, the Philippines and Japan, where steam has long been essential both for maintaining schedules and coping with the volume of traffic.

The Mediterranean, home to famous sailors from at least the days of Ulysses, if not longer, was also slow to adapt to steam, at a time when the countries along its shores were undergoing a long period of secular decline. At the end of the eighteenth century the Mediterranean still retained, from the sixteenth century, some of its institutions – described in such detail by Fernand Braudel's classic studies[19] – among them the use of slaves as oarsmen for galleys. Sailing ships manned by Corsairs – essentially pirates from the Barbary Coast of North Africa – were still attacking ships at sea, not only in the Mediterranean but sometimes as far away as the English Channel, in order to enslave their crews. Their operations lay behind President Thomas Jefferson's decision, in 1798, to found the United States Navy, which only five years later, in 1803, landed marines in Tripoli, a Corsair stronghold, to rescue captive American citizens. His motives were largely financial; since, after 1783 Britain would no longer ransom captive American citizens, this unwelcome burden fell on the new United States. After Tripoli, American marines went on to capture Algiers in 1812, but in the nineteenth century it was France that put an end to the Corsairs' operations – simply by occupying the coast of North Africa from Morocco to Tunisia. The galleys – like the triremes of ancient Greece – were warships, best suited to narrow seas, with light changeable winds and few good natural harbours. In Europe they operated not only in the Mediterranean but in the Baltic, where, on 15 May 1790, both sides fought with galleys in the decisive Battle of Fredrikshamn in the Gulf of Finland, which brought victory to Sweden at the end of a two-year war with

Russia. In point of time, however, galleys, by the age of steam, had almost disappeared from the world maritime scene – together with the institutions, such as slavery, which they supported; they left no important legacy. Piracy, needless to say, survived in other forms. As for other shipping in the Mediterranean, it had, by the end of the eighteenth century, few distinctive forms: by this time, commercial traffic through the straits of Gibraltar and into the North Atlantic had long been the order of the day.

4

THE INVENTION OF THE STEAM ENGINE AND THE PROBLEM OF LOCOMOTION

Today's concept of the standard steam engine, operated by high-pressure steam driving a piston within an enclosed cylinder, is based upon inventions that date back no further than the end of the eighteenth century. Until then, steam engines – in widespread use, throughout the century, for a number of different purposes – operated according to a quite different principle, which was derived from the scientific discovery of atmospheric pressure by the Italian scientist, Evangelista Torricelli (1608–47). Following this discovery, a number of experiments carried out by noted European and British scientists confirmed the power of a vacuum to sustain a remarkably heavy weight, such as that of a 28-foot-high column of water. Practical applications depended upon developing pumps that would create a vacuum by removing air from an enclosed space. Somewhere around 1650, a German burgomaster, Otto von Guericke, using an air pump of his own invention, showed how a vacuum created within a hollow copper ball would make it crumple up 'as a cloth is crushed between the fingers'.[1] Using a more robust container, consisting of a piston closely fitted into an air-tight cylinder, with the outside end of its rod attached to a system of pulleys, he showed how the piston, at the top of its stroke, could support a weight of 2,868 lb. The noted British scientist, Robert Boyle, consistently improved the air pump in order to carry out more experiments of this kind – publishing the results in 1669. At the same time the Dutch scientist, Christiaan Huygens – who like Boyle came from an extremely wealthy family – pursued the same line of research in Paris, where he worked in the laboratory of the recently founded Académie Royale des Sciences. There he employed a young Huguenot assistant, Denis Papin, who, fearing religious persecution, left Paris for London, where Boyle, in 1675, took him on as an assistant in his experiments with air pumps. There he conceived of the idea that if the vacuum in a cylinder such as that used by von Guericke could be created within a much shorter period of time than could be achieved by a state-of-the-art air pump, then the action of the piston could provide the basis for a working engine.

Papin first worked with gunpowder, which, on exploding, consumes oxygen. The apparatus failed for simply being too inefficient – on account of the quantity of air still left in it after the explosion. Papin, however, also saw that the same result could be achieved, if somewhat less dramatically, by condensing steam. His new apparatus consisted of a water-tight vertical cylinder, in which the piston, after a downward stroke, would rest on a small quantity of water. Then, after a fire was lit underneath the cylinder, the water would boil to yield, inside a minute, a much greater volume of steam, whose pressure would force the piston to the top of the cylinder. As the cylinder cooled, and the steam condensed into water, the resulting vacuum would provide the force for the downward stroke of the piston..

Although Papin's apparatus was far short of providing the basis for a practical working engine, by the time he died – probably in 1712 – an Englishman, Thomas Newcomen, had invented a working steam engine based upon the same essential principles. In Newcomen's engine, however, a separate boiler produced steam to be fed into an adjacent cylinder containing a large piston. How much, if anything, Newcomen's engine owed to Papin, is uncertain. A man living in Cornwall could well have missed the review of Papin's invention published in the *Philosophical Transactions of the Royal Society* for 1697.

The industry that gained the greatest benefit from Newcomen's steam engine was coal-mining. Throughout the seventeenth century mines were abandoned, not because the coal seams were exhausted but because of flooding. Up to a certain level, underground water could be contained, in one way or another, but beyond this level lack of sufficient power to pump it out – whether provided by men or horses – meant that the battle was lost. Waterwheels and even windmills were occasionally used for driving pumping machinery, but they were only marginally useful. The coal, however abundant it might be, stayed in the ground.

The first engine designed by Newcomen was installed to pump water out of the coal mine at Dudley Castle in Staffordshire in 1712. It had the great advantage of being self-acting, in that all the water-cocks and valves required by the hydraulic system were opened and closed by a system of rods driven by the beam. The capital investment must have been considerable, given the vast scale of the whole structure. The brick building housing the engine was more than 40 feet high, and the length of the beam was some 25 feet.[2] The engine operated at a rate of up to sixteen 6-foot strokes a minute.[3]

The unprecedented efficiency of the Newcomen engine in pumping water, from depths of up to 300 feet, was almost immediately recognized by mine-owners as justifying the considerable capital and operating costs. Within three years of 1712 engines, on a scale determined by the depth and volume of the water to be pumped, were being constructed throughout England – one of the earliest, brought into use in 1715, being at Tanfield Lea, one of the coal mines served by the spectacular wagonway described on page 29. In 1722 the first Newcomen engine on the Continent was installed at a mine in Slovakia, and by the time Newcomen died in 1729 his engines were also to be found in Hungary, France, Belgium and possibly Germany and Spain.[4] The engines were not always used for pumping out mines: one built at Passy, then just outside Paris, was used to pump water for the city from the Seine, at the stupendous rate of nearly ten million gallons a day.[5]

In practice, however, the vast rate at which the engines burnt coal meant that it was hardly economical to operate them except at coal mines, where, *a fortiori*, coal was abundant. Moreover, there was always a sufficient stock of unmarketable 'small coal' – which accumulated at every pithead – to fuel the Newcomen engine. The bottom line did require, however, a gain in productivity sufficient to cover the substantial maintenance fees paid to the engineers, to say nothing of capital costs and patent rights. Even so the rapid spread of Newcomen engines in the coal-mining world testifies to the way they transformed it.

One result was a vast increase both in the number of wagonways and in their volume of traffic. In Tyneside this led to the development of double tracks, with the main way serving the loaded wagons descending by gravity to the staithe and the bye-way providing the track for horses to pull the empty wagons back to the mine. Significantly the earliest known case dates from 1712, the same year as Newcomen introduced his steam engine. The old single-track system with occasional passing loops, known as 'sidings', was still sufficient for the majority of mines. What is more, a double-track system was only workable when there were no uphill gradients on the main way, a result achieved, at considerable expense, by the Tanfield wagonway. This was, however, far the busiest of all the wagonways, carrying about half the Tyneside traffic and a third of the total traffic of the north-east coalfields. Wagons ran at an average interval of 45 seconds, a rate four times higher than the average for all wagonways.[6] Generally, once the new system was accepted, nothing essential required the bye-way always to keep close to the main way. This was important, since the main way could work with a gradient of 1 in 10. Albany's Run on the Tanfield wagonway was 1 in 12 for a quarter of a mile, and the Bryans Leap Run on the Western Way, which dropped 500 feet in a mile, must have been almost as terrifying as its name suggests.[7] On the other hand, horses pulling empty wagons uphill on bye-ways could not manage anything steeper than 1 in 30. (Downhill, on the main way, the horse simply trotted behind the wagon – to which it was attached by a loose rope and halter – as can be seen from any number of contemporary illustrations.[8] It was not part of its task to prevent the wagon running away out of control.)

The vast increase in production made possible by the Newcomen engine, and the considerable expansion of the mining infrastructure that followed, coincided with the gradual substitution of coal for charcoal in iron-smelting as the result of a process developed by Abraham Darby and introduced by him at Coalbrookdale in Shropshire in about 1713. Darby's process was used to produce cast iron, which within a generation was adopted for the manufacture of the cylinders and boilers in Newcomen engines, replacing brass and copper – which were much more expensive. According to some self-professed experts, this could have been a false economy, since the walls of a brass cylinder could be no more than one third of an inch thick, while with cast iron they could never be less than one inch. Given that the cylinder was alternately heated and cooled with every stroke, up and down, of the piston, the difference in thickness was critical to the consumption of fuel; even so the cheapness of coal at the pithead, combined with that of iron in relation to brass, meant that when it came to the bottom line the former won against the latter. One economic consequence was an increase in the dimensions of the cylinder, so that in the second half of the eighteenth century, one with a diameter of 70 inch was a standard size. The largest cylinder, recorded in 1769, had a diameter of 75 inches.[9] Its length would have been more than 10 feet and its weight more than 6 tons.[10] The work required in transporting such a vast object from the foundry to its destination at the pithead represented, needless to say, a considerable part of its cost. At this level more than one boiler was needed to provide enough steam to fill it, and more than one jet of water to condense the steam. All this was obviously a considerable boost to the iron industry, at the same time as it provided an incentive for better grades of coal (including coke) to be produced for it. On both sides the demands for transport naturally increased very substantially.

Although, at the end of the day, improved design and workmanship, particularly on the part of John Smeaton (1724–92), had doubled the efficiency of the Newcomen engine, its limitations – after half a century of widespread use – became more difficult to accept every day. At the pithead its usefulness was confined to pumping out water, contributing nothing to the work of winding coal up the shaft – the power driving the winding drum was still supplied by horses. At a time, also, when the limitations of waterwheels as the main source of power in the textile industry were becoming equally unacceptable, there was room for something better than the Newcomen engine.

The decisive shortcoming was that the engine was purely reciprocal: if it could move a beam up and down, it could not – in any efficient way – make a wheel rotate. (At the very end of the eighteenth century an atmospheric engine was made rotative by using a connecting rod, crankshaft and flywheel, but such machines – known as 'whimsies' – were so clumsy that their usefulness never went further than winding coal from shallow pits.[11]) If, in principle, the familiar foot-lathe and spinning wheel provided a model for translating reciprocal into rotatory motion, it offered no solution to the problem of adapting it to the scale of a Newcomen steam engine. To start with, this engine was only powered on the downstroke of the piston. There was in any case a crying need for something better: the Newcomen engine was so wasteful of fuel that it was hardly economical to install it at Cornish tin mines, where the problem of flooding was just as acute as in coal mines along the Tyne and Wear.

Almost all the many shortcomings of the Newcomen engine were overcome as a result of one of history's most remarkable business partnerships – that between the inventor, James Watt, and the entrepreneur, Matthew Boulton. In 1769 Watt was granted a patent for 'a new invented method of lessening the Consumption of Steam and Fuel in Fire Engines': his new atmospheric engine, with a separate condenser, represented an improvement that would so greatly increase its efficiency as to make its use economical even when coal had to be imported over a considerable distance – such as that between Tyneside and Cornwall.

Watt had made his invention in his home town of Glasgow. Although, by profession a surveyor and instrument maker, it was as an indefatigable inventor that he had noted the shortcomings of the Newcomen engine. His invention of the separate condenser led to his meeting Dr John Roebuck, a local captain of industry, who also owned leases of a coalfield and blast furnaces from the Duke of Hamilton. Roebuck, impressed with Watt's invention and seeing its potential utility in his own enterprises, proposed a business partnership to exploit it; Watt accepted this, but first – in July 1768 – Roebuck required him to go to London to arrange the legal procedure that led to the 1769 patent. Returning to Glasgow, he made a detour to Birmingham to meet Boulton – with an introduction from Roebuck.

Watt and Boulton took to each other immediately – so much so that Watt stayed for two weeks at Boulton's house in Snow Hill. By this time Boulton was the owner of a factory at Soho in Staffordshire, where the 600 craftsmen he employed made products of such taste and workmanship that the name of Soho was known worldwide. Returning to Glasgow Watt did his best to persuade Roebuck that Boulton would be an ideal business partner, but Roebuck would go no further than offer Boulton a licence to manufacture the engine in the 'Midland Counties only'. Boulton's reaction to this offer, after it had been communicated to him in a letter from Watt, can be seen from the final sentence of his reply: 'It would not be worth my while to make for three counties only, but I find it very well worth my while to make for all the world.'[12]

Boulton was wise to bide his time. In 1773, Roebuck, heavily in debt as a result of poor yields from his coal mines, had no choice but to make a composition with his creditors, who included Watt. Since the other creditors would not give a farthing for Watt's new engine, Roebuck's two-thirds share in the 1769 patent was Watt's for the asking. This left him free to negotiate anew with Boulton, but the patent, with only eight years to run, was only interesting to Boulton if it could be renewed for a much longer period. Watt presented the necessary bill to Parliament on 23 February 1775, and with the help of lobbying by Boulton, it overcame 'violent opposition from many of the most powerful people in the house',[13] to receive the Royal Assent on 22 May. The Act, which would also apply to Scotland, extended the patent for twenty-five years. A partnership contract agreed on 1 June provided that, for the whole of this period, Boulton would be entitled to two thirds, and Watt, one third, of the property in the patent, with the former paying all expenses, and the latter being obliged 'to make drawings, give directions and make surveys'[14] at an agreed remuneration of £300 a year. This was a good bargain for both sides.

Such was the beginning of a remarkable business partnership. Within a year two vast new engines were installed, one at the Bloomfield Colliery in Staffordshire, and the other, to blow air, at John Wilkinson's Brosely Ironworks in Shropshire – an application with a considerable future, given the growing industrial use of iron and steel. This was just the beginning: the Bloomfield engine was the model for many others throughout the United Kingdom, and from 1779 in France also. There was some local resistance in Cornwall, where the new engines were certain to save costs in pumping out water from the tin mines, which – for hundreds of years – had a near monopoly throughout almost the whole of Europe;[15] the 'no cure, no pay' terms offered by Boulton and Watt – based on one third of the savings in fuel over a period of twenty-five years – saved the day.

Success with what, after all, was still a reciprocating engine, was not enough for Boulton, who rightly saw that new industry – particularly in textiles – offered a profitable market for a rotatory engine. Until the end of the eighteenth century, the power required by a mill was supplied by either wind or water, and of these only the latter could meet the demands of the heaviest machinery – and then only by being stretched to the limits of its capacity. The watermill put a premium on industrial locations close to a significant rapid fall in levels on natural waterways. The power then available depended both on the height of the fall and the volume of water. Inevitably textile mills concentrated in regions – such as are typical of the north of England – with fast-flowing streams coming down from the hills. In such places watermills were often part of the local scene.

As a source of power for winding engines, a rotatory engine would be equally useful for mining: cages could bring much heavier loads up the shafts, and, at a later stage, empty wagons could be hauled uphill back to the pithead.[16]

A number of problems had to be solved before the power of a reciprocal engine would be translated into a rotating wheel, and it was only after much persuasion by Boulton that Watt proceeded further.[17] Although spinning wheels and lathes, powered by the reciprocating action of a foot-treadle, pointed the way to a solution – with a flywheel driven by a crank coupled to the beam of a reciprocating engine – Watt was for some time concerned that variations in the length of the piston stroke would cause trouble: while this did not matter when the working beam's only function was to operate a pump, it would frustrate the successful operation of a crank-drive.

Watt's concern was true to his character as a born pessimist. He need not have worried: the rotation of the flywheel would itself ensure a piston stroke of constant length – the tail would wag the dog. With hindsight it is surprising that Watt did not see this; his dithering did, however, allow a Birmingham business rival, James Pickard, to obtain a patent, in 1780, for a crank-drive, so that Boulton and Watt had to find an alternative to cover the fourteen-year period until expiration of the patent in 1794. (Watt was himself convinced that he had first conceived of the crank-drive, and that an employee, Ned Ruston, had leaked the idea to Pickard, in which case the latter's patent could well have been challenged in court.) Watt, by temperament, preferred to invent alternative mechanisms. Within a year he had invented no less than four different alternative mechanisms, all covered by a patent of 1781. Of these only the sun and planet gear worked satisfactorily. Its operation meant that the flywheel made two complete rotations for every cycle of the engine; this was not necessarily a disadvantage, and Boulton and Watt continued to use it occasionally even after Pickard's patent expired in 1794.

A much greater problem lay in the fact that steam engines were still single-acting – that is, only the downward stroke of the piston was power-driven. The alternative was to design a double-acting engine, where the upward stroke of the piston was also power-driven. As early as 1775 Watt conceived of an enclosed cylinder in which steam would be admitted not only below the piston-head, to power the downward stroke, but above it, to power the upward stroke. In 1782 a double-acting engine was included in an omnibus patent granted to Watt. Significantly the patent also covered the application of the new engine to wheeled carriages. This was done mainly at the instigation of a young, but highly valued assistant, William Murdock (who was engrossed by this particular problem), but Boulton and Watt also wanted to be protected against pre-emption by others experimenting in this field.

As so often happens in life, solving one problem only led to another. With a single-acting engine, because a chain, which always hung vertically, linked the piston and the working beam, there was no horizontal force on the piston rod. This was inherent in the design of the two ends of the working beam, both of which took the form of an arc of a circle whose centre was the axis of the beam. Since a chain – like the rope of an old-fashioned bell-pull – can only 'pull', not 'push', it is of no use for the link between the piston and the working beam of a double-acting engine. With a single rigid strut, however, the thrust arising from the up-and-down motion of the beam, if not counteracted, would subject the joint with the piston rod to horizontal forces, which in turn would impede its free vertical motion. This may not sound like much of a problem, but Watt rightly saw it both as a threat to the smooth running of his engine and as a source of unnecessary wear and tear. He solved the problem by introducing a pantograph frame. The 'parallel motion' achieved by this arrangement was always the invention of which he was most proud.[18] It was first incorporated in the No. 1 engine of the Albion Flour Mill opened in London in 1788, in which each pair of millstones – there were ten in all – produced an unprecedented 10 bushels of flour per hour. The mill burnt down in 1791, allegedly as the result of arson. The vast economies of scale had enabled flour to be sold at prices far lower than could be offered by any standard wind- or watermill, and this certainly gave established millers – who made no secret of their hostility to the new mill – every reason to rejoice.

Although – with the exception of the use of rotatory engines to hoist skips up mineshafts – Boulton and Watt made no direct contribution to the development of steam-powered locomotion, the rapid introduction of their engines in factories created a demand for transport of coal and other raw materials which – even with the vast expansion of the British inland waterway network at the end of the eighteenth century – became steadily more difficult to satisfy. Not surprisingly, many potential inventors – on both sides of the Atlantic – contemplated the prospect of adapting the steam engine to locomotion. Indeed, as shown by the steam-driven mechanisms already in use for hauling both wagons and boats uphill by means of a cable wound round a drum connected to an engine,[19] the first step had already been taken. (This system is still the most efficient in special cases, such as funicular railways.) The breakthrough would come when the engine itself became part of the transport – so representing the ultimate in locomotion. Only when this was achieved would the limitations on distance – inherent in all cable systems – be overcome.

The problem was that the atmospheric engine, even with all the improvements resulting from Watt's inventiveness, was never going to be suitable for mounting on a vehicle that would travel overland. This turned largely on the limitations of a dedicated track, even when it was one with iron rather than wooden rails; its dimensions could simply not be extended to cope with a wagon carrying any existing model of an atmospheric engine. The best any such wagon might achieve would be to move itself along a level railway – a result actually achieved experimentally before the end of the eighteenth century. What was needed was a locomotive with a surplus of power sufficient to pull a train of wagons – itself a recent innovation in short-distance transport from coal mines.

As the size to which ships could be built was quite disproportionate in relation to anything possible with land transport, there was nothing to prevent a rotatory atmospheric engine being installed as a source of power. The pioneer in this field was the French Marquis Claude Jouffroy d'Abbans who, in June 1778, made an unsuccessful attempt with a steam-powered boat on the River Doubs: five years later, he did succeed, with his *Pyroscaphe*, which, on 15 July 1783, made its way for fifteen minutes upstream on the River Saône. In France, revolution soon blocked further progress, leaving the field to American enterprise. On 1 February 1788, the state of Georgia – for the only time in his history – issued a steamboat patent to Isaac Briggs and William Longstreet, the first ever granted in the US. Two years later, in 1790, John Fitch established an infrequent but continuing service on the Delaware river, linking Philadelphia in Pennsylvania to Trenton in New Jersey. This – 'the first successful and practical application of steam to ship propulsion'[20] – was followed before the end of the century by similar successes on the Connecticut and the Hudson. Steam navigation as a practical undertaking, with a reliable service and fixed timetables, had to wait for Robert Fulton's *Clermont*, commissioned by Robert Livingston, who, in 1798, had been granted a monopoly on steam navigation in the waters of New York, provided he could operate a successful steamboat by 1805. Although the *Clermont* (then known only as the 'North River Steamboat') only entered service in 1807, Livingstone's monopoly still held until, in February 1824, it was overturned by the US Supreme Court.[21] Every advance in water transport, from Fitch to Fulton and beyond, was calculated to meet the traffic demands on the great rivers of the American east coast, which – as already noted in Chapter 3 – provided abundant incentive for improving on wind and sail.

When it came to the technology of steam, Fulton was hardly an innovator. The *Clermont* had a 20-horsepower James Watt-type engine driving a paddlewheel crankshaft. It was a flat-bottomed 100-ton ship – designed for both passengers and freight – with which Fulton, a businessman as much as an inventor, operated pleasure trips on the Hudson River between New York and Albany. A contemporary called her 'a monster moving on the waters, defying wind and tide, and breathing flames and smoke'. Fuelled by pinewood rather than coal, the ship moved faster than any steamboat before, attaining a speed of 5 miles an hour and travelling 130 miles on its maiden voyage. Within three months, it had earned $1,000 against the initial cost of$20,000.

Long before the end of Livingston's New York monopoly in 1824, the development of high-pressure steam engines foreshadowed the end of the era of Watt and Boulton's atmospheric engines. The new high-pressure engines, essential for steam locomotion over land, also proved to be superior when it came to marine transport. Watt himself had long known of their potential, but – cautious as ever – was inhibited by the danger in working with high-pressure steam. History was often to prove him right: in the new steam age of locomotion, exploding boilers were an ever-present hazard.

What then did high-pressure steam add up to, and what were its advantages? The basic principle was simpler than that of the atmospheric engine. Provided its pressure is sufficiently high, steam admitted into an enclosed cylinder containing a piston, will overcome any force to drive the piston head. The problem is to find a boiler strong enough to withstand steam pressure at a level at which it can provide useful mechanical power, given the dimensions of the whole set-up. Once this level is reached, a valve mechanism operated by the engine can be designed to admit steam at the beginning of every stroke; close to the end of the stroke this valve is closed, and another opens to release the steam, which escapes rapidly into the open air. A mechanism can combine the valves in a single unit. In the simplest form of single-acting engine, with an upright cylinder, the piston, having reached the top of its stroke, then falls under its own weight, to open once again – at the bottom of the stroke – the valve which admits high-pressure steam from the boiler. This completes a cycle that can be repeated indefinitely.

The principle readily extends to enable a double-acting engine, with the piston in a totally enclosed cylinder, so that steam can both be admitted and allowed to escape on both sides of the piston head. Since, with such a system, the piston returns to the beginning of the cycle under the pressure of steam, there is no need for the cylinder to be vertical – the more so, since, in contrast to any atmospheric engine, there is no need for a horizontal working beam. With a horizontal cylinder – the optimal position for certain purposes – gravity will not be a factor in the action of the piston. A slightly more complicated system of connecting rods and tappets is to be seen in the high-pressure engine and boiler built to a design of Robert Trevithick – the leading inventor in this field,[22] but the principle is the same. In this case the vertical cylinder is hardly visible behind the large horizontal boiler with its conspicuous chimney.

The high-pressure engine is a landmark in the history of steam, if only for its simplicity and versatility. Its merits are well summed up in H.W. Dickinson's classic *A Short History of the Steam Engine*: [23]

> It is difficult at the present day to realize what fundamental changes were involved in introducing the high-pressure engine. The ponderous beam see-sawing slowly in a massive engine house crowded with valve gear, air pump and condenser, costly to install and run, was replaced by a faster-running direct-acting engine with a simple valve gear, occupying little space, requiring hardly any foundation, cheap in first cost and easy to work.

Sometime around the turn of the nineteenth century, Trevithick, seeing all these favourable attributes in the engines he was making, realized that they could be adapted to power a vehicle on a common road. He arranged to present his prototype to the public on Beacon Hill, just outside his home town of Camborne in Cornwall, on Boxing Day, 1801. The presentation, and one following it on 27 December, was hardly a success, largely because the party which gathered to admire Trevithick's carriage made too much of the festivities appropriate to the season.[24] Nonetheless, his friends were impressed, and urged him to go to London and obtain a patent. On 24 March 1802, Trevithick was granted Patent No. 2599, entitled 'Steam engines – Improvements in the construction thereof and Application thereof for driving carriages'.

Success came at last on 21 February 1804, not on a road but on a cast-iron tramway. In South Wales, where Trevithick's stationary engines were much in demand for powering rolling mills at ironworks, he put up one of his best clients, Samuel Homfray, to bet a neighbouring ironmaster, Anthony Hill – a friend as well as a rival – 500 guineas that he could produce a locomotive to run on the plate-way then linking both their ironworks to the Glamorganshire Canal. (A plate-way – an alternative to a railway – consisted of flat plates, some 6 to 7 inches wide, supported longitudinally on wooden sleepers: the wagons, or 'trams', that ran on it were specially constructed for this type of track, which was favoured by many mine-owners and ironmasters throughout Britain.[25]) Within a few weeks Trevithick had produced a locomotive that ran successfully, over a distance of nearly 10 miles, from Homfray's ironworks to Navigation House on the Canal. Not surprisingly, there was some damage to the track, and Trevithick's locomotive had the advantage that it was downhill all the way. Although 1804 is a year to be remembered in the history of steam locomotion, Trevithick was too busy building and marketing his stationary engines – not only in Britain but also in such faraway countries as Peru – to proceed with development of locomotives. (He did, however, produce a locomotive named 'Catch Me Who Can' for an exhibition in London in 1808, where it ran round a circular track.) Even so, he was rightly described, two generations after the triumph of 1804, as

> the real inventor of the locomotive. He was the first to prove the sufficiency of the adhesion of the wheels to the rails for all purposes of traction on lines of ordinary gradient, the first to make the return flue boiler, the first to use the steam jet in the chimney and the first to couple all the wheels to the engine.[26]

The problem of adhesion, which continued to frustrate the development of practical locomotion for many years at the beginning of the nineteenth century, was founded on the misconception that the level of friction between metal wheels and metal rails was too low for the system to operate. The wheels, powered by the steam engine, would spin rather than drive the locomotive forward. Although Trevithick's various locomotives would appear to have proved the contrary, their achievements were disregarded as special cases, not generally applicable. Therefore a solution was found for what was actually no problem at all. This was to have traction provided by a rack-and-pinion drive, with a saw-tooth rack laid along the rails being engaged by the teeth of a pinion, which, in turn, was powered by the steam engine. The inventor of this system, John Blenkinsop, built two locomotives, the 'Prince Regent'[27] and the 'Salamanca',[28] for use on a railway in Yorkshire linking Leeds to a coal mine at Middleton; these two locomotives, inaugurated on 12 August 1812, were the first ever to provide a regular service.

Although this unwieldy arrangement held its own for a number of years, it was never in widespread use. Later on it came into its own on mountain railways, where it still operates – as, for example, in the Bernese Oberland, on the electric railway between Interlaken and the Jungfraujoch – at 11, 300 feet the highest railway station in Europe. The Blenkinsop locomotive was, however, introduced on the Kenton and Coxlodge tramway, to be used for coal traffic on north Tyneside. There its operation attracted the interest of a sceptical observer, Christopher Blackett, owner of the nearby Wylam Colliery, who doubted the necessity of the whole rack-and-pinion system. He proved his point by building a wagon, to run on state-of-the-art iron rails, with windlasses attached by gearing to several wheels. Six men then climbed onto the wagon to operate the system, and as soon as they set to work the wagon moved along its track without slipping. After further trials Blackett was satisfied than an engine, conformable with the wagon, could be produced with sufficient power not only for the wagon itself, but also for a whole train of wagons behind it.

Blackett's problem was that his Wylam tramway was a plate- rather than a rail-way. The difference between plate and rail was critical when it came to being able to carry the weight of a steam engine: plates, whatever expedients – such as the 'Dilly', a locomotive with eight wheels constructed at Wylam – were adopted to spread the load, still could not support it. Even one day's operation threatened to crack a plate somewhere along the line. The owners of a neighbouring mine at Killingworth, impressed by what had been achieved at Wylam, believed that their engine-wright, George Stephenson, could go one better and produce a viable locomotive – taking advantage of the fact that their mine worked with railways. Stephenson, employing skills laboriously acquired by working with steam engines, and at the same time being up to date with the latest technology as represented by Wylam's 'Dilly', produced a locomotive for Killingworth, which had its first trial run on 25 July 1814. Named 'Blücher',[29] and running on Killingworth's cast-iron rails, it overcame the defects of all its predecessors in steam locomotion – so much so that one admirer, in 1821, noted that 'the locomotive engine of Mr. Stephenson is superior beyond all comparison to all other Engines I have ever seen.'[30]

George Stephenson, soon known to history as 'the Father of the locomotive',[31] was a man of humble origins but with a remarkable gift of perseverance. Every day, as a boy growing up in a poor Tyneside coal-mining family, he saw wagons drawn by horses and running on wooden rails – the classic scenario already described in Chapter 2. Inevitably he too went into mining, where his earliest ambition was to become an engineman. Starting as a fireman, he rose, after three years, to be a 'plug-man', responsible for the operation and repair of a pump engine in Killingworth. Then aged eighteen, he realized that he would go no further without learning to read and write, so he took lessons (also in arithmetic) three nights a week after work.

Once into his twenties he soon gained the reputation of being able to make repairs to machinery where others had been defeated, in one case making it possible to reopen one of the Killingworth mines after it had been flooded for a year. At the age of twenty-eight he was appointed engine-wright to the mine, so becoming responsible for all the steam-powered machinery. From almost his first day his innovations substantially reduced working costs – an achievement much appreciated by his employers. It is not surprising, then, that they entrusted him, in 1814, with the construction of a locomotive to haul wagons on the railway down to the River Tyne – where, by this time, the rails were made of iron, not wood. The 'Blücher' was the result. Its success, which was immediate, was the result both of better design and better machining – Stephenson, having seen all too often how mining machinery failed for poor workmanship, set unprecedentedly high standards of precision. At the same time he was continually improving every part of his locomotive – the engine itself, the wheels (made from malleable instead of cast iron) and the springing (essential on the uneven tracks) all received attention. The same was true of the rails, together with the chairs and sleepers that supported them. Although Stephenson's reputation for thoroughness soon led to commissions to lay down new railways throughout Tyneside, these were all private undertakings following the established pattern of moving coal from mines down to the river. If use of the steam locomotive had gone no further, the efficiency and productivity of British coal-mining would still have been considerably enhanced. This, however, would hardly have added up to a transport revolution. A new perspective on the potential of railways with steam locomotives was needed. This was also discovered in the coal-mining region of north-east England, and realized in practical form with the opening of the Stockton and Darlington Railway on 27 September 1825. The new era that then opened is explored in Chapter 8.

5

THE AMERICAN RIVERBOAT

When I was a boy, there was but one permanent ambition among my comrades in my village on the west bank of the Mississippi River. That was to be a steamboatman.

From Mark Twain, *Life on the Mississippi*[1]

Down the Missippi (*sic*) steamed the Whipperwill,
commanded by that pilot, Mister Steamboat Bill.
The owners gave him orders on the strick Q.T.,
to try and beat the record of the "Robert E. Lee."
Just feed up your fires, let the old smoke roll,
Burn up all your cargo if you run out of coal.
If we don't beat that record, Billy told the mate,
"send mail in care of Peter to the Golden gate."

The wife of Mister William was at home in bed,
When she got the telegram that Steamboat's dead.
Says she to the children, "Bless each honey lamb,
the next papa that you will have will be a railroad man."

From Ren Shields, *Steamboat Bill* (1868–1913)

It is not only for its romance, but also for its importance in economic history, that the adaptation of steam power to shipping in inland waterways is an essential theme in this book. It is, however, one to which Britain, in spite of the achievements of such men as James Watt and George Stephenson, contributed relatively little. These, and many others, took it for granted that a horse – or sometimes a team of men – walking along a tow-path provided the most economical power for boats on inland waterways. On occasion a steam-powered bank-engine, winding a towline on its vast drum, could be useful – as was also the case with railways – but this was never going to be the power-house of an integrated transport system. The great age of British canals – roughly speaking the last quarter of the eighteenth century[2] – came a generation too early for any account to be taken of the advantages of steam power for boats. This meant that canals were built on too small a scale for navigation by boats carrying the weight of a steam-engine.[3] At the time there was an obvious cost-advantage, and when, in the nineteenth century, the advantages of building on a larger scale became apparent, competition by railways soon showed that such investment was unlikely to pay dividends. This was not seen as a shortcoming. The British canal network transformed the carriage of freight at a time when the new factories, born out of the Industrial Revolution, had to be supplied with raw materials and fuel – mainly coal – on a quite unprecedented scale. If the capacity of canal boats, by the standards of the mid-nineteenth century, were small, and their rate of progress determined by that of a horse walking along a tow-path, these factors hardly counted when the canals were built. It would be a mistake to suggest that their construction was unnecessary. The canals were an indispensable component of the economic structure of the early Industrial Revolution.

When it came to transport on inland waterways, the need for a new source of power arose not from the use of canals, but rivers. There is a critical difference between the two. Canals, being man-made, constitute a network whose form is determined by purely economic considerations. They are built to the standard dimensions of the boats that will use them and normal hazards of river transport – such as water levels that can vary drastically in response to natural forces, navigable channels too far to be reached from any tow-path, powerful currents that make progress upstream problematic – are avoided. However, rivers were generally moderate in their rate of flow, and relatively free of hazards. Excesses of nature, such as flooding, could be allowed for on the principle that 'what cannot be cured, must be endured'. Setbacks created by natural forces counted for little when set against the usefulness of river transport. This was particularly true during the course of the eighteenth century, when the demand for coal produced by the mines close to such rivers as the Tyne and Wear gradually extended across the whole country to reach such distant consumers as the Cornish tin mines with their steam-driven pumps.[4] It was largely for such traffic that boats powered by steam were first introduced in England.

For the real success of the steamboat on inland waterways one must turn away from Britain. In the first half of the nineteenth century, which was critical for the development of inland steam navigation, the United States led the field. There, the early development of steamships for traffic in the rivers and along the coasts of the eastern seaboard has already been noted in Chapter 4. Although this process was already underway by the end of the eighteenth century, the centre of action in the nineteenth century would be far inland, in the vast area drained by the Mississippi and its tributaries of which nine would give their names to states admitted to the Union in the sixty-six years from 1792 to 1858.[5] Equal in importance were the five Great Lakes, Ontario, Eire, Huron, Michigan and Superior, which – except for Michigan – were shared with Canada. Although as early as 1679 the French built the first sailing ship, the *Griffin*, on the upper Great Lakes, the slow pace of exploitation up until almost the end of the eighteenth century meant that the age of sail counted for very little in either one of these two great American systems of inland waterways. Their eventual exploitation – a key factor in the economic history of North America in the nineteenth century – depended almost entirely on vessels powered by steam.

The interaction between the British, the French and the Indian tribes is critical for understanding the historical background. Starting at the mid-point of the eighteenth century, the British had by this time established the thirteen colonies along the Atlantic Coast that, by the end of the century, would constitute the United States. Already, by the end of the seventeenth century French explorers from Canada – for more than a hundred years a French colony based on Quebec – had advanced westwards to discover all the five Great Lakes, which geographically constitute a vast inland sea comprising the world's largest area of fresh water.

Access, however, was extremely problematic: the altitude of Lake Ontario, although linked to Montreal by the upper St Lawrence River, was 243 feet above the highest point of the river that could be reached by ocean-going vessels. Beyond Montreal the Lachine Rapids made navigation next to impossible: the substantial difference in altitude was critical. The waters of Lake Erie, at an altitude of 603 feet, after flowing out along the Niagara river, fell some 360 feet at the falls before reaching Lake Ontario. The usefulness of the lakes was

MISSISSIPPI MAIN WATERWAYS
CANADA
Lake Itasca
Yellowstone
Snake
Minneapolis
Missouri
Rocky Mountains
Platte
Omaha
Illinois
UNITED STATES
Ohio
Appalachian Mountains
Colorado
Arkansas
St Louis
Cumberland
Memphis
N
Red
Mississippi
Key
International frontier
State boundary
River
New Orleans
Rio Grande
0
500 km
0
311 miles
MEXICO
Gulf of Mexico

Map 1

further limited by the fact they were frozen during the winter, so although the French had few rivals for well over a century, they profited little from their vast domain.

The French were still intrepid explorers, and in 1681, the Sieur de Lasalle, having crossed the frozen lakes in mid-winter to a point close to where Chicago was later established, crossed overland – with Indian guides – to reach the frozen Illinois river, a tributary of the Mississippi. Bringing their canoes downstream on sledges to what is now the town of Peoria, de Lasalle and his party met open water. They soon reached the Mississippi and by the early summer of 1682 they were on the shores of the Gulf of Mexico. On paper, at least, de Lasalle had acquired a vast new empire for the King of France.[6] Beyond quite intensive fur-trapping, the French were in no position to exploit it. Then, by the Treaty of Paris at the end of the Seven Years' War (1756–63) – in which rival British, French and Spanish claims in North America were a major issue – France lost almost all its territory east of the Mississippi to Britain, while the rest, including everything west of the river, went to Spain. Given that the London-based Hudson's Bay Company had become the owner of much of what is now the west of Canada in 1671, the British, in one way or another, became the owners of much the greater part of North America in 1763. At this stage it hardly mattered that Spain had acquired Louisiana – as the French had called their territory west of the Mississippi. For London what counted was that the French were no longer serious rivals in North America.

By this time the citizens of the Atlantic colonies, who were mainly English, German and Dutch, were already moving westwards to seek their fortunes in and beyond the Appalachians. Indeed it was largely their encounters with the French, far inland, that had led to the Seven Years' War. In spite of the obstacle represented by the mountains, they were much closer to their home base on the eastern seaboard than the French were to Quebec. What is more, they were not only much more numerous, but they also had the Indian tribes of the Iroquois on their side.

The territory lost by France east of the Mississippi was added to that of Britain's thirteen Atlantic colonies, who were soon at odds as to how it should be divided up between them – a situation that was still unresolved at the end of the revolutionary war and the recognition of the independence of the United States in 1783. In 1803, as an incident in the Napoleonic wars, the Louisiana territory was recovered by France; three weeks later, the United States, under President Thomas Jefferson, bought it all for some $15 million.

By this time tens of thousands of settlers from the east had claimed land beyond in the Appalachians. Already in 1792, the admission of Kentucky as the fifteenth state, had extended the American frontier to the Mississippi, with Tennessee following on in 1796; in the first quarter of the nineteenth century, they were joined by four states included – wholly or in part – in the Louisiana purchase, Alabama (1816), Louisiana (1812), Mississippi (1817) and Missouri (1821). Cotton worked by slaves, and producing a commodity destined mainly for export to Europe, was the base of their economy. All this reflects a breathtaking pace of settlement and development, with next to nothing in the way of a transport infrastructure to support it beyond trails – long known to local Indian populations – through the forests of the Appalachians.

At much the same time as American pioneer settlers were moving west beyond the mountains, others, who had chosen to remain loyal to the British crown, moved north to Canada. There, in 1791, the British had established, as Upper Canada, the vast territory, hardly settled by the French, bounded to the north by the Ottawa river and to the south by the St Lawrence River, and the northern shores of the Great Lakes. The continuing British presence in Upper Canada was perceived as a threat by the new United States, and it took two more years of war (1812–14) to resolve the line of the frontier. The Treaty of Ghent in 1814 recognized the United States as entitled to all the territory south of the Great Lakes, including the whole of the present state of Michigan – which was essentially the position before the war started. Upper Canada then corresponded to the present Canadian province of Ontario, while Lower Canada, now the province of Quebec, remained the home of the greater part of the French-speaking population.

As a result of all these developments, the Americans west of the Appalachians and the British in Upper Canada faced essentially the same problem, that is, to develop a vast region which enjoyed both considerable if unrealized natural wealth – in furs, lumber and minerals – and an even greater potential for agriculture and industry. Yet these settlers faced the same problem as in the end had defeated the French: in spite of the considerable advantages – long recognized in Europe – of transport by water over transport by land, the Great Lakes and the Mississippi river system did not constitute a viable communications infrastructure for the United States as it was at the turn of the nineteenth century. Even so, in the first half of the nineteenth century, the economic demands of the new states east of the Mississippi could only be met by exploiting the potential of both systems, and in the case of the Mississippi this would have been impossible without the steamboat.

Almost inevitably, the success of steamboat operations along the American east coast, and above all on the great rivers flowing into the Atlantic (described in Chapter 4), led those involved to turn their sights to the Mississippi, and more particularly, its great eastern tributary system based on the Ohio river. What is more, from the whole character of these rivers it was clear that there was no practical alternative to steam when it came to propelling boats upstream. Robert Fulton, encouraged by his success on the Hudson, combined with Robert Livingston (who had played a key role in negotiating the Louisiana purchase) to exploit the much greater potential of the Mississippi. They had, however, an immediate setback, because both Ohio and Kentucky refused to follow the precedent established by New York state and grant them a monopoly similar to that which they enjoyed on the Hudson. Nothing daunted, they went ahead and in Pittsburgh built the *New Orleans*, a steamboat on the model of the *Clermont*. Leaving on its maiden voyage in October 1811, it arrived, ten weeks later, in January 1812, at New Orleans. Fulton and Livingston then decided to use it only on the short run between New Orleans and Natchez, some 200 miles upstream on the east bank of the Mississippi, where it was lost to fire in July 1814. Their decision was purely commercial: upriver from Natchez to Louisville on the Ohio, 'a river distance of 1,000 miles, there was only a frontier wilderness with a thinly scattered population of backwoodsmen and no towns of importance to supply traffic and support to steamboats'.[7] In any case, with its hull uncomfortably deep for the Mississippi's frequent shallow waters, the design of the *New Orleans* was ill suited to the upper river. What is more, Louisiana, only admitted as a state in 1812, did grant the monopoly refused by Kentucky and Ohio, although Fulton and Livingston only enjoyed it for three years, for both died in 1815.

By this time an actor much better attuned to the geography west of the Appalachians held centre stage. In 1807, Henry

Shreve, who, having grown up in Brownesville, Pennsylvania on the banks of the Monongahela – which, at Pittsburg, joined with the Allegheny to form the Ohio river – had much better first-hand knowledge of the Mississippi and its tributaries than Robert Fulton, began to build a boat which took into account their wayward character. This he took successfully down the Ohio to the Mississippi at Cairo, Illinois, where he turned upriver intent on loading a cargo from the Indian lead mines on the Galena river – a small eastern tributary which joins the Mississippi in Illinois, just south of the Wisconsin state line – far to the north of Cairo. The Indians, used to selling to the French and English who came down the Mississippi from Lake Superior in pirogues, were at first reluctant to sell to a temperance man who, instead of hard liquor, could only offer household goods such as pots and pans, but in the end they finally traded their entire hoard of some sixty tons of lead to Shreve. This he loaded onto a flat boat specially constructed under his own supervision in situ, which he towed downriver the whole way to New Orleans. There he transhipped to a schooner, which he sailed to Philadelphia where he sold the whole cargo at a profit of $11,000.

Shreve was not only the first man to take a steamboat up the upper Mississippi, but also the first, in 1814, to captain a boat upriver the whole way from New Orleans back to Brownesville – an achievement remarkable enough to be noted by the East Coast press. This boat, the *Enterprise*, which only just made it, was not designed by Shreve, who knew only too well that his success with such a deep draught was only possible because the Ohio river had been in flood.

Shreve once again realized that he had to build his own boat. This had a flat-bottomed hull 136 feet long and 28 feet wide, which was too shallow for machinery. Instead Shreve decked it over, and put the engine and boilers on deck. This he covered with a second deck, and above this he placed the pilot-house, with two high smokestacks just behind it. These were for the two separate, efficient, lightweight, high-pressure engines,[8] each with its own horizontal boiler and piston for powering one of the side-wheels. The *Washington*, a boat built on this design at Wheeling, on the upper Ohio, left the city on its maiden voyage on 4 June 1816, with Shreve as its captain.

Still on the Ohio, only a few hundred miles downriver, the *Washington* ran aground, and as the crew tried to pull her off with an anchor, one of its two boilers blew up, throwing both crew and passengers, including Shreve, into the water. Several died and others suffered excruciating pain, in what was the first, but far from the last accident of this kind. Shreve, recovering almost immediately, got his act together, buried the dead, repaired the boat and steamed the rest of the way down to New Orleans without further incident. Once there he confronted the monopoly granted to Fulton and Livingston, but following their deaths in 1815, there was little will to enforce it, and it effectively lapsed in 1818.[9] What was really remarkable, not only in New Orleans but along the whole river, was Shreve's ability to turn the boat round and steam back to Louisville on the Ohio, in a record twenty-four days. In spite of Shreve's earlier success with the *Enterprise*, it was this achievement that removed all doubts and prejudices about the future of steam navigation, so encouraging the building of shipyards in every convenient locality.[10] The *Washington*, however, was much the largest boat on the river, which explains why its success captured the public imagination.

Shreve had designed, built and captained the classic Mississippi paddle steamer – a model that remained standard for generations. Western steamboats, following his basic design, differed very substantially from the model, which

Fulton had brought from New York. With little resistance to wear and tear they were short lived and extravagant with fuel, but their high-pressure engines were cheap to build; they carried freight as much as passengers, who had to accept a safety record quite horrifying by today's standards. Boiler explosions and fires on boats mainly built of wood could destroy both cargoes and the lives of passengers in a matter of minutes. This was a different world from east of the Appalachians, where fast, comfortable, relatively safe and long-lived passenger boats, with low-pressure engines – built at much greater cost – dominated the relatively low volume of traffic. By the 1850s – by which time the voyage upriver took as few as four days – Shreve's steamboats 'outweighed in tonnage all the vessels of the Atlantic seaboard and the Great Lakes combined'.[11] They were the most notable achievement of the American industrial revolution. By 1840, when the steamboat – which had started thirty years before as a haphazard, unskilled, local improvisation[12] – had reached its most perfect form, with engines rated on average at 3½ times the power of those used by industry on land, it accounted for three fifths of all the steam power used in the United States.[13] In the 1850s New Orleans overtook New York in volume of shipping, with half of all American exports moving through the port. By this time also, the economic development of the Mississippi and the Ohio extended along the whole length of both rivers, with St Louis – in the mid nineteenth century the fourth largest city in the US – located at the point where the Missouri joins the Mississippi, as its hub. It is no wonder then that the river steamboat was admired and copied worldwide as far afield as Siberia, Latin America, Egypt and the Congo.

By 1840, however, the American transport infrastructure was being transformed by the construction of railroads, which inevitably affected steamboat traffic on the Mississippi river system. In the mid nineteenth century river and rail traffic both complemented and competed with each other, with the terms on which they did so defined mainly by economic factors. As a general geographical statement, the bias of the Mississippi river system was north–south while that of the railroads was east–west. In the early days before the Civil War (1861–65) – a watershed in American history – river traffic dominated: steamboats had had, after all, a quarter of a century start on railroads, and it was only in 1852 that the Pacific Railroad Company became the first to complete a line west of the Mississippi. Now part of today's Missouri–Pacific Railroad, this was intended to be the first link in a line to the Pacific coast. Before the Civil War it never extended beyond the state of Missouri, where its main function was to bring passengers to the start of the western wagon trails.

What then did the Mississippi system add up to in its heyday? One of the great pioneers of western expansion, Senator Thomas Hart Benton (1782–1858) of Missouri – which in 1821 had been admitted as the first state wholly west of the Mississippi – reckoned that some 50,000 miles of water in the Mississippi system were navigable by some kind of boat; in any case some 16,000 miles of steamboat routes are recorded.[14] They differed considerably both in the character of their waterways and in that of the traffic they carried. For convenience they are best assigned to four different categories, defined by the geography of the system. The character of the Mississippi from St Paul, Minnesota, the highest point of the navigable river, changes radically at St Louis, Missouri, where the Missouri river flows in from the west. The course of the river from St Louis, down to the Gulf of Mexico, some hundred miles downstream, defines the southern section – with New Orleans as the transhipment point between river and sea traffic – whereas the upper river, above

St Louis, defines the northern section, navigable up to St Paul in Minnesota. The Ohio river, and its main tributaries, defines the eastern section, leaving the western section to be defined by the Missouri and its tributaries.

From 1811, the year of the maiden voyage of the *New Orleans*, steam soon took over on all four sections – not that there had ever been much traffic in earlier years. The steamboats, with their high-pressure engines, burnt the wood, which was just as abundant as in the east in the natural forests along the river banks. Any number of shipyards constructed them according to Shreve's basic design: they were cheap to build and unlikely to be in service much longer than five years given the hazardous conditions of the waterways for which they were produced in considerable numbers. Their short lives had any number of possible causes. The rivers were full of snags, which in early days were mainly the branches of trees that had fallen into the water as a result of erosion of the river banks, but later on, in increasing measure, the debris left by wrecks. In addition, cargoes, notably cotton, were highly inflammable, and cheap boilers – only too likely to explode under the pressure of steam – took any number of lives, causing untold damage not only to the boat itself but also to its surroundings. The fact that in the first thirty years of the steamboat era 185 boiler explosions resulted in more than 2,000 fatalities gives some idea of the level of danger.[15] In 1824, hundreds – including many onshore – died on the lower Mississippi when the steamboat *Clipper* exploded, but little was done to make the riverboats safer. In 1838 the US Congress – after President Martin van Buren had twice highlighted the problem in his State of the Union address – passed the Steamboat Bill, but effective legislation had to wait until 1852; by then the reckless pioneering days were passed.

Steamboat races were another Mississippi river spectacular, as suggested by the lines quoted from Ren Shields's *Steamboat Bill* at the head of this chapter. The steamboat *Whipperwill* seems to be the product of poetic licence, but the *Robert E. Lee* was the winner of the most famous steamboat race of all, held in June 1870, between New Orleans and St Louis – and commemorated in a Currier and Ives coloured etching. The competing boat was the *Natchez VI*, which had already made a record-breaking trip from New Orleans to St Louis in 3 days, 21 hours and 58 minutes. Even so the *Robert E. Lee* won the race, but only after the *Natchez VI* had suffered the misfortune of being stuck on a mudflat for six hours. Its captain, T.P. Leathers, might have avoided this fate – and won the race – if, like Captain John W. Cannon of the *Robert E. Lee*, he had stripped his boat of excess weight and declined any passengers or cargo. In any case the *Natchez VI* did not suffer the fate of the *Whipperwill* as suggested by the poem.

The Mississippi steamboats were manned by a motley crew. On every one of them there were five men who counted: first was the captain, who, if he had the same authority as his ocean-going counterpart, was essentially a businessman, if not the actual owner, whose task it was to run the business, at the same time keeping the cabin-class passengers – many of them his clients – happy. Next came the pilot, who from the wheel-house, had the best view of the river. From this vantage point he had to find the right passages, both up and downstream, in a river that constantly changed its character – both in time and place. In navigating the boat the word of the pilot – described by Mark Twain as 'the only unfettered and entirely independent human being that lived in the earth'[16] – was law.

The engineer, who was responsible for keeping the engines going, had a much lower place than the pilot in the shipboard hierarchy. Where a pilot only got his licence after serving a long and demanding apprenticeship, almost anyone was acceptable as an engineer – which almost certainly explains the frequent and devastating boiler explosions. The engineer did not do his work alone: firemen, working four-hour shifts, continually fed wood into the voracious boilers.

In practice the clerk ranked in status next to the captain, at much the same level as the pilot. He was the business manager and accountant, an exacting task when both passengers and cargo were continually being taken on board, made no easier by the fact that fees were paid in a variety of notes issued by banks up and down the river. A clerk accepted most of these only at a discount, which made his task even more complicated.[17]

The lowest ranking officer on the boat was the mate, but his work – as a labour foreman – was still indispensable. The labour force consisted mainly of roustabouts, 'the proletariat of the river population'.[18] Working twenty-four hours a day, with little respite, the roustabouts, with little mechanical help, loaded and unloaded cargo, and just as important stored it in such a way that the boat remained stable. When the steamboat grounded on a shoal – an all-too-common event on the Mississippi – their task was to tranship the cargo onto flatboats to lighten the steamer. On top of all this the roustabouts had, twice a day, to carry 4 foot logs on board as fuel for the engines. Most of them were Irish or German immigrants, who, after landing at New Orleans were recruited almost immediately for this arduous and dangerous work, which cost many of them their lives. Any who fell into the river were simply left behind, as were also, on occasion, hapless lower-deck passengers.

The classic Mississippi river scene, with lower-deck passengers crowded in among bales of cotton, was to be found along the 1,200-mile-long southern section of the river between St Louis and New Orleans, where there were 1,327 landings at which boats might stop.[19] Much of the cotton was brought down to the main river along its mainly shallow tributaries such as the Arkansas to the west and the Tennessee to the east. Captains, keen on business, were always ready to pull in to collect or discharge cargo, even if it meant losing time on scheduled runs. This scene, with modifications according to the character of the passengers and cargo carried, was repeated in other sections of the river where there was much less traffic, particularly in the early days before the canal, and later rail links, to the east, were opened; here cotton, essentially an export crop destined mainly for the mills of Lancashire, gave way to other cargoes, including, particularly along the Ohio, coal from the mines of western Pennsylvania and Virginia, and farm produce from the new lands settled by immigrants from the east,[20] and along the upper river and the Missouri, furs and minerals. The upper river, also, was a major source of hardwood, with the logs floated downstream adding to the hazards of navigation.

As for passengers, almost no facilities were provided for the lower-deck ones, who, in their hundreds, paid next to nothing for their transport. Particularly on the lower river even their lives counted for little – on one occasion reported by the Memphis *Daily Eagle*, a hundred German immigrants were put ashore in freezing weather on an island, to lighten the boat so that it could cross a bar. The boat then sailed on, leaving them to their fate.[21] In contrast the facilities provided for the cabin passengers, who paid ten times as much, were luxurious by local American frontier standards, even if not seen as such by the occasional sophisticated easterner or European traveller – such as John James Audubon[22] and Charles Dickens.

The saddest of all the steamboat passengers on the lower Mississippi were Indians from the eastern states, transported under the federal Indian Removal Act of 1830, which required all belonging to this category to be moved west of the Mississippi. In 1831 the law was brutally enforced by President Andrew Jackson, with Indians belonging to the eastern tribes brought down to the lower Mississippi and then transported by steamboat up its two main western tributaries, the Arkansas and the Red rivers, both of which were navigable for several hundred miles; their destination was a vast area comprising the whole of the present state of Oklahoma, and parts of Kansas and Nebraska, formally set aside in 1834 as Indian territory.[23] In 1838 the forced migration culminated in the Trail of Tears, the name given by history to the removal of 16,000 Cherokees from their ancestral lands in Georgia, escorted by 7,000 soldiers; that some 4,000 Cherokees died en route gives some idea of the conditions under which they travelled. The transport of Indians in the 1830s was, however, exceptional: the steamboat services on the Arkansas and Red rivers were developed mainly for transporting soldiers and supplies to the forts established by the US Army in the process of opening up the west to settlers. The same was true of the western tributaries, including, above all, the Missouri, of the upper river. In 1876 the process reached a dramatic climax on the Little Big Horn, a distant tributary, via the Yellowstone River, of the Missouri, where General George Custer, in command of some 260 US Cavalry, was defeated by a coalition of Plains Indians led by Crazy Horse and Sitting Bull. Significantly the whole operation – of which Custer's battle was but one incident – was conceived for the benefit, not of shipping, but of railroad interests concerned to construct a northern route to the Pacific coast.

Back on the old Mississippi, the lower-deck passengers were a very mixed bunch, but particularly on the Ohio and the Missouri many were settlers from the east seeking new land. From the early 1840s onwards steamboats on the Missouri, after leaving St Louis, brought countless settlers to Independence, on the other side of the state, the starting point of the Santa Fe and the Oregon trails, whence they continued their westward journey in the covered wagons familiar from many a Hollywood epic. In the mid 1850s, however, almost the whole of Kansas (and much of Nebraska to the north of it) was the scene of a bitter strife between settlers from the southern states, who wished to establish a 'slave' state, and those from the northern states who with equal fervour wished Kansas to be 'free'. In the end, its prospective admission to the Union as a free state in 1861 was a major factor leading to the secession of eleven southern states and the Civil War that then followed.[24] If the fighting in Kansas did not discourage the overlanders with their covered wagons, it certainly delayed, until after the end of the Civil War in 1865, the construction of railroads across the continent to the Pacific coast.[25]

Indeed the Civil War radically changed the whole Mississippi steamboat scenario. Because the line-up in the war was basically North-South, steamboat operations on the lower river were controlled by the Southern states of Arkansas, Louisiana, Mississippi and Tennessee, while those on the low-upper river, together with the Missouri and the Ohio, were controlled by some nine Northern states.[26] Here however the situation was complicated by the fact that Virginia, whose capital, Richmond, was also the Southern Confederate capital, included some 300 odd miles of the left bank of the Ohio river. This hilly and densely forested part of the state belonged geographically to Appalachia, and had been settled by northerners who had little in common with the tidewater[27] plantation owners who – totally dependent on slave labour – had ensured that their state declared for the

Confederacy. The effective result was that the North controlled both banks of the Ohio.[28]

By the time of the Civil War the steamboat services on the upper Mississippi and the Ohio had become part of an integrated communications infrastructure, which served a rapidly growing industrial and mining economy. By this time the fact that the war, with the blockade of the lower river by the Confederate states, denied access to the sea via the lower Mississippi and the port of New Orleans mattered relatively little to the northern states. The links with the east coast established by the canals and railroads built in the generation before the war made such access unnecessary.

On the other side, because the main rivers of the southern states east of the Mississippi, such as the Cumberland and the Tennessee, were tributaries, not of the Mississippi, but of the Ohio, the Confederacy suffered grievously from the northern control of the river. What is more, early in 1862, the Union General Ulysses S. Grant, captured two key Confederate forts, Fort Henry on the Tennessee and Fort Donelson on the Cumberland. The fall of Fort Henry followed Grant's victory at nearby Shiloh in what was not only the greatest battle of the Civil War, but also one of the most decisive in favour of the Union. By this time Confederate attempts to capture the key city of Louisville on the Ohio had failed.

From the beginning of the war the economy of key cities on the lower Mississippi, such as Memphis, Tennessee, and above all, New Orleans, was seriously affected by the loss of all traffic from the upper river. Worse still, the export of cotton – essential to the southern wartime economy – was hit hard by the Union's naval blockade of the Gulf and Atlantic coastlines. Worse was to come. The lower Mississippi offered an obvious line of attack for the Union forces, and in 1862, Admiral Farragut, attacking from the sea, captured New

Orleans, while to the north Union forces moving both down-river and overland captured Memphis. This left only some 200 miles of the Mississippi, with Vicksburg as the mid-point, in Confederate hands. At every stage of the river war the Union was served well by the fact that few of the shipyards that built the steamboats were on the lower Mississippi: in the supply of both labour and materials – particularly the metals needed for the engines – the North had always been far ahead of the South. When it came to defending Vicksburg, the Confederate forces did succeed in improvising a number of remarkable men-of-war, most with bales of cotton instead of armour plating. These cotton-clads, helped by one or two of the earliest ever ironclad warships, fought hard to defend the river at Vicksburg in the first battles in history to be fought between armoured steamships. In the end the Union forces – which often fought ineptly – were overwhelming, and Vicksburg finally fell on 4 July 1863; the whole of the Mississippi was then in Northern hands, so cutting off the states of Arkansas and Louisiana from the rest of the Confederacy. The victory was as important for the Union cause as the one – much better known to history – at Gettysburg, east of the Appalachians, a day earlier. In neither case did the Confederate forces ever come close to recovering the lost ground: their final defeat in 1865 had become inevitable.

After the end of the war recovery by the southern states, at least according to their own perception, was made doubly difficult by the abolition of slavery. (It also counted that, in the four years of war, England's Lancashire textile industry had found alternative sources of cotton, notably Egypt and India). With peace, little time was lost in extending the railroad network west of the Mississippi, as related in Chapter 9: the classic era of the American river steamboat had ended.

Given that the transport links essential to the continental economy which followed from opening up the west, ran mainly east–west – as they still do – railroads were the only answer.

It is now time to turn back the pages of history, and examine how it came about that the Mississippi river system, and more particularly the upper river and the Ohio, had, by the time of the Civil War, been integrated into the communications infrastructure east of the Mississippi – a development that, before the end of the nineteenth century, would ring down the curtain on the classic steamboat scenario familiar to any reader of Mark Twain. Until 1825, the only link between the Mississippi river system and the outside world was by sea, via the leading harbour city of New Orleans. Even before the turn of the century this had been seen as an intolerable situation, particularly by the men on the frontier, whether in the United States or Upper Canada. The answer, at least on paper, could be found simply by looking at a map. The extensive natural inland water systems had to be connected to the flourishing economy of the eastern seaboard; this in turn depended on a number of major ports, long linked to each other by coastal shipping, which – in the first quarter of the nineteenth century – was rapidly converting to steam, as was also true of the shipping on the long, broad and mainly navigable rivers flowing into the Atlantic. Of these, the Hudson, flowing almost due south through eastern New York state, to enter the sea at New York City, was economically the most important. What is more, its western tributary, the Mohawk, rose close to the shores of Ontario, the most eastern of the Great Lakes.

In 1792, with the support of General Philip Schuyler, the Federalist leader in the New York State Senate, the Western Inland Lock Navigation Company was incorporated to open a navigable waterway from Albany – where the Mohawk joined the Hudson – to Lake Ontario. The company's strategy was two-fold: first, the navigation of the Mohawk would be improved – mainly by the construction of locks and short lengths of canal – to accommodate the latest Durham boats, 60 feet long and displacing sixteen tons; second, the Oswego river, that flowed into Lake Ontario, would be linked to the upper Mohawk by a new canal. Although the first objective was achieved, the improvements to navigation of the Mohawk failed to make it a viable traffic artery. As to the second, the Oswego canal never really got started. Even if successful, the plan had a fatal defect: Lake Ontario, separated from the four upper lakes by the Niagara Falls, provided no access to them. Americans in upper New York State and Canadians in Upper Canada might have benefited from the link being completed, but even then Montreal – linked to the lake by partial canalization of the Lachine rapids of the St Lawrence – could have proved to be a better gate to the outside world than New York.[29] This possibility proved to be a powerful argument in favour of the direct link to Lake Erie, which was what the new economy west of the mountains demanded.

In 1810, at a meeting of the promoters of the Western Inland Lock Navigation Company convened to find a solution to its problems, a state senator, Jonas Platt, asked 'Why not make application at once for a canal to connect the waters of Lake Erie with the waters of the Hudson River?' To Platt's surprise his suggestion was immediately taken up by the other promoters; the problem was to obtain the necessary government approval. The solution found by a predominantly Federalist company was to persuade a prominent and influential political opponent, the Republican De Witt Clinton, to support an official resolution to appoint 'a board of commissioners to examine and survey the entire interior route from the Hudson to Lake Erie as well as that to Lake

Ontario and around the Niagara Falls, and to report on the most eligible path'.[30] With Clinton's support it was agreed unanimously by the state legislature.

The 1810 resolution was only the first step in a political process that would not be resolved until 1817, when, after many ups and downs in the intervening years – which included two years, 1812–14, of war with Britain – Clinton, on 1 July, finally became governor of New York state, having been elected on a platform that plainly promised the construction of the Erie Canal. At a ceremony on 4 July, Independence Day, the first soil was broken, and the whole canal was completed in 1825. This was, to quote from the standard history:

> an outstanding engineering feat. A narrow ribbon of water 363 miles long, 40 feet wide at the top, and 4 feet deep was created between Albany and Lake Erie ... In overcoming the 565 feet of elevation of Lake Erie over the Hudson at Albany, the Erie Canal followed a combined ascent and descent of 675 feet. The canal had eighty-three locks with lifts ranging from six to twelve feet, and a succession of eighteen aqueducts which became the marvels of the day.[31]

In the two-year period before the official opening of the Erie Canal on 26 October 1825, the greater part of the canal had been open to traffic – so much so that in the final month 8,000 men, 9,000 horses and 2,000 boats were employed in the transportation of goods. The boast, made in one of the speeches at the opening ceremony, that New York had 'made the longest canal – in the least time – with the least experience – for the least money – and of the greatest public utility of any other in the world', was justified.[32] At a time when British and continental canal traffic was at its peak the scale of the enterprise far exceeded that of any European rival.

The whole concept of the Erie Canal was quite different: boats were ready to carry not only any freight offered but also passengers, for whom purpose-built 'packets' were designed. The passenger journey between Albany and Buffalo took only four days, a remarkable time given the length of the canal. Travel by coach on New York's turnpikes was faster, but much more expensive. The main limit on canal traffic was defined by the minimum 4 feet depth of water and the standard dimensions – 90 × 15 feet – of the locks; the rise on the other hand varied from one lock to another according to the demands of the local topography. In practice, canal users, intent on enjoying maximum capacity, built their boats so that they just fitted into the locks.

The question to be answered now is simply what place have canals, on which all boats were towed by horses or mules, in a history of the age of steam. A simple question demands a simple answer. The fact that the natural waterways of North America, whether rivers or lakes, were already being taken over by steamboats in the years preceding the opening of the Erie Canal,[33] was the strongest argument in favour of its construction, even though no steamboats would operate along its length. Notwithstanding the costs of transhipment, its usefulness as an integral part of the communications infrastructure was immeasurable. The figures speak for themselves when it comes to looking at the tolls earned by the canal in the years after 1825 in the light of the original capital invested.[34] What is more, there was generally no need to tranship goods from the canal at the Albany basin: the canal boats, from fifty to a hundred at a time, were lashed together and towed down the Hudson by steamboat. By 1846 fifteen purpose-built steamboats fulfilled this function, in addition to another hundred or more 'unequalled in any part of the world for speed and accommodations' carrying passengers and freight.[35]

In the early years after 1825 De Witt Clinton's enthusiasm for the Erie Canal was soon shared by many others, who attributed the supremacy of New York City over rival Atlantic seaports to the commerce it brought. The hype continued for a long time, so that as late as 1845 a newspaper editor could claim that:

> The settlement of Western New-York and Ohio forced the construction of the Erie Canal, which literally united the waters of the western seas with the Atlantic ocean.[36] For only twenty years, the wealth of the teeming West has poured down that avenue, and already it has placed New-York on an eminence as the Commercial Emporium of America ... So long as New-York remains at the head of the western trade ... it must irresistibly advance in wealth, influence and population, until she will be known not only as the great city of America, but as the great city of the world.[37]

By this time, indeed, the demands of traffic had led to a considerable increase in the dimensions of the canal and its locks.[38]

In spite of the many different sorts of steamboat on the Hudson, they almost all shared two common properties: except for the actual steam engine, and the transmission to the paddle wheels, the boats were not only made of wood, but burnt wood as fuel – just like the Mississippi steamboats. Given the abundant forest cover of the east of the United States – and Canada also – this intensive consumption of wood made good sense economically, at least in the short term. In the end, steamboats switched to coal – just as the railroad did[39] – but the bituminous coal best suited for their engines was to be found in western Pennsylvania on the far side of the Appalachians,[40] so that bringing it to the east coast and its great rivers, such as the Hudson, had to wait upon the construction of canal links to the Ohio river, and its tributary, the Allegheny.

In constructing such links there were three separate strands of development. The first was to link the Ohio to the Great Lakes. Here the southern shore of Lake Erie was once again the obvious site for canal terminals. Where Buffalo, at the eastern end of the lakes, had been established as the terminal for the canal link to the east coast via the Hudson river, Cleveland and Toledo, respectively some 200 and 300 odd miles west along the lake's southern shore, became the terminals for two canal systems linked to the Ohio river. In both cases the location depended on a navigable river flowing into the lake, the Cuyahoga in the case of Cleveland, and the Maumee in that of Toledo at the far western end of Lake Erie. The year 1832 witnessed the opening of the so-called Erie and Ohio Canal, which, with the Cuyahoga river providing the first stage from Cleveland, joined the Ohio at Portsmouth. The Miami and Erie Canal – named after the Miami river that accounted for much of its length – linking Toledo with Cincinnati was opened in 1845.

The second strand consisted of constructing a direct link between the Great Lakes and the upper Mississippi. The seventeenth-century explorer, Louis Joliet – the first European ever to see the upper river – had already conceived of a canal along the line of the short portage between a sluggish stream called 'Checagou' by the local Indians and the little Des Plaines river. The former flowed into Lake Michigan and the latter, via the Illinois river, was a tributary of the Mississippi. In the whole complex network of American canals there was none that was more clearly justified on the score of economic potential, and as early as 1814 President James Madison asked the United States Congress for the necessary authorization. The request failed for lack of interest even though by this time shallow-draught Mississippi steamboats were already reaching La Salle, the town proposed as the canal terminal on the Illinois River, while on Lake Michigan they were reaching the mouth of the Chicago, where in 1825 the eponymous city was made an incorporated town of the second class. In spite of the fact that only 50 miles of canal would need to be dug across country much easier than that surmounted by the Lake Erie canals, the Illinois and Michigan Canal was only completed in 1848 – by which time Chicago already had some 20,000 inhabitants.

The third strand of development was the construction of a link between the Ohio and the Potomac, so opening the whole Mississippi river system to Atlantic traffic.[41] If the great barrier of the Appalachians always stood in the way of such a link, the construction of the Erie Canal proved that it could be surmounted. Just as the city of Albany on the Hudson river, at the head of the Mohawk river, was the obvious terminal for such a canal, Washington, at the end of the Potomac tidewater, was indicated as the eastern terminal of any canal link to the Ohio. What is more, the course of the Potomac above Washington – where it was hardly navigable – indicated the route to be followed by the eastern part of the canal. The Potomac was also the frontier between the states of Virginia and Maryland, and the final section of the canal in Pennsylvania – which in the event was never built – would link Cumberland, in the mountains of western Maryland, with Pittsburgh on the Ohio.

The task of building the canal was assumed by the Chesapeake and Ohio Canal Company which was organized at a meeting of prospective stockholders in Washington in June 1828.[42] Two weeks later formal inauguration, with the US President John Quincy Adams turning the first spadeful of earth, took place, appropriately, on 4 July, and by the end of August contracts for constructing different sections of the canal were agreed with a number of firms with previous experience in New York, Pennsylvania, Connecticut, Ohio and Canada. Construction, however, was very slow for the first five years, the reason being that the canal company faced opposition of a kind unprecedented in the history of the United States.

Although, in the year 1827, the world's only public railroad operating with steam locomotives was England's famous Stockton & Darlington Railway, in that same year a group of American merchants in Baltimore met to organize the Baltimore and Ohio Railway Company. Its purpose, which was essentially the same as that of the Chesapeake and Ohio Canal Company, was to provide a transport link between Chesapeake Bay – on which Baltimore, then the United States's second city, was the major port for sea-going shipping – and the Ohio river. The possibility that the canal would lead to a new port – on the Potomac close to Washington – that, with the advantage of the direct link to the Ohio, would challenge the supremacy of Baltimore was incentive enough for the merchants of that city to promote a railroad. Their problem was that for most of the way to the Ohio the valley of the Potomac was the only viable route. Up-river from Point of Rocks, a Maryland township some 40 miles north-west of Washington, there were any number of points where there was hardly room for both a canal and a railroad.

The result, which was in the interests neither of the canal nor the railway factions, was a critical loss of construction time during some five years of litigation and political lobbying, allowing 'New York and Philadelphia [to forge] ahead as commercial centres at the eastern end of a rich and growing western trade via improved transportation systems'.[43] The canal never went further than Cumberland in the mountains of western Maryland, and final completion had to wait until 1850. In 1852, only two years later, the railroad did reach the Ohio, at Wheeling, Virginia. In doing so it had to cross the Buckhorn Wall, a natural feature at an altitude of 2,500 feet in the heart of the mountains, which would have been an almost insurmountable obstacle to canal-builders. As related in Chapter 7 the Baltimore and Ohio Railroad, although now part of the extensive CSX system, played a major part in American railroad history. From Harper's Ferry and many other points, long freight trains are still to be seen moving slowly along its tracks in the valley of the Potomac. If the canal, on the other hand, has become little more than a favourite route for hikers, in its heyday it played a major part in developing the Cumberland coalfields. As conceived of in 1828 it was to be built on a scale greater than that of the Erie Canal, but its promotors failed to foresee the hard game played by the City of Baltimore and what it regarded as its own railroad company.

In the end the other canals that opened up the vast lands in the new states beyond the Appalachians were also doomed. Long-distance transport that required transhipment to vessels drawn by horses was never going to survive the challenge of the new railroads. By 1850 it was becoming clear that modern canals must be built on a scale to accommodate river – and even ocean-going – steamboats. Such were the Chicago Drainage Canal and the New York State Barge Canal on which construction started at the end of the nineteenth century. The former, opened in 1900, supplanted the Illinois and Michigan Canal, while the latter – whose final sections were opened only in 1918 – supplanted the Erie Canal.

Although construction was protracted over more than two decades the New York State Barge Canal – ten times as long as the Panama Canal, which was opened four years earlier – was in the early twentieth century as remarkable an achievement as the old Erie Canal had been a hundred years beforehand. The greater part of the modern barge canal was created by adapting the natural river system to the standards required by shipping. The long section from Troy (at tide-water level) to Rome (a summit at 420 feet) consists of a canalization of the Mohawk river, while the short 24-mile-long link to Lake Ontario consists mainly of the Oswego river.[44] In this case, the canalization by Germany of the Upper Rhine, at the end of the nineteenth century, already showed the vast potential of this expedient.[45] Its drawbacks would only become apparent at a later stage.

6

THE STEAMBOAT AND INLAND WATERWAYS

> It was a great comfort to turn ... to my influential friend, the battered, twisted, ruined, tin-pot steamboat. I clambered on board. She rang under my feet like an empty Huntley & Palmer biscuit tin kicked along a gutter; she was nothing so solid in make, and rather less pretty in shape, but I had expended enough hard work on her to make me love her. No influential friend could have served me better. She had given me a chance to come out a bit – to find out what I could do.
>
> From Joseph Conrad, *The Heart of Darkness*[1]

In the history of steamboats on inland waterways, Britain, in spite of the achievements of its great engineers such as James Watt and George Stephenson, counts for little. These, and many others, took it for granted that a horse – or sometimes a team of men – walking along a towpath were all that was needed. On occasion a steam-powered bank-engine, winding a towline on its vast drum, could be useful – as was also the case with railways – but this was never going to be the powerhouse of an integrated transport system. The great age of British canals – roughly speaking the last quarter of the eighteenth century[2] – came a generation too early for any account to be taken of the advantages of steam power for boats. This meant that canals were built on too small a scale for navigation by boats carrying the weight of a steam engine – so that, for example, the middle stretch of the Grand Trunk Canal, linking the Trent and Mersey rivers systems, could only accommodate boats with a beam not greater than 7 feet, if they were to be able to pass each other; this established the narrowboat standard that is still in force today.[3] At the time the established system had an obvious cost advantage, and when, in the nineteenth century, the advantages of building on a larger scale became apparent, competition by railways soon showed that such investment was unlikely to pay dividends. Long before the end of the nineteenth century Britain was thus left with a largely obsolete canal network. Rivers were important only in so far as they adapted to ocean-going steamships, which explains, above all, the ascendancy of two relatively short rivers, the Mersey and the Clyde, in the nineteenth century. The importance of the Thames, where steamboat traffic emerged at a very early stage, had long been established.

On the principle that necessity is the mother of invention, two different types of vessel, characteristic of such rivers, called for the introduction of steam power. These were, on one side tugs, and, on the other dredgers. In both cases successful operation requires the vessel to be able to follow an exact course in a small area, in which the demands of navigation are measured in feet rather than miles. A tug, whose function is to bring a much larger ocean-going ship to the right harbour anchorage, must meet these demands, which by the mid eighteenth century had become a major concern both of the Royal Navy and the British mercantile fleet.[4] Although designs for the use of steam power appeared as early as the 1730s, success only came in 1802, with its use in tugs on the Forth–Clyde Canal.

A dredger must be able to position itself precisely above any shallow reach of water where the river or seabed must be excavated to create a deeper channel for ocean-going vessels. The dredger is almost unique in requiring power for two purposes: the first is bringing the vessel itself to the area to be dredged, while the second is operating the actual system installed for excavation. As early as 1561, a ladder dredger, with power supplied by horses, operated on the Brussels–Schelde canal. The use of a continuous chain of scoops, excavating the underlying soil and bringing it to the surface to be unloaded into the hold of the dredger, cried out for the application of steam power: this was first achieved in 1803 when a steam-dredger, designed and built by Robert Trevithick, was used to dig out London's East India Dock.

Of all the world's navigable rivers none – from the beginning of the age of steam – could rival the Rhine for its usefulness to the economies of the countries – Switzerland, France, Germany and the Netherlands – through which it flows. The river itself rises in eastern Switzerland; since the mid nineteenth century, but not before, navigation has been more or less unimpeded along its whole length from the famous falls at Schaffhausen to the North Sea in the Netherlands. The cities along the Rhine – Basel, Strasbourg, Cologne, Arnhem and many others – reflect its importance both over centuries of history and at the present day. As shown by the map on page 105, its tributaries, both to the east in Germany, and to the west in France, greatly extend its reach – in much the same way as those of the Mississippi on the other side of the Atlantic.

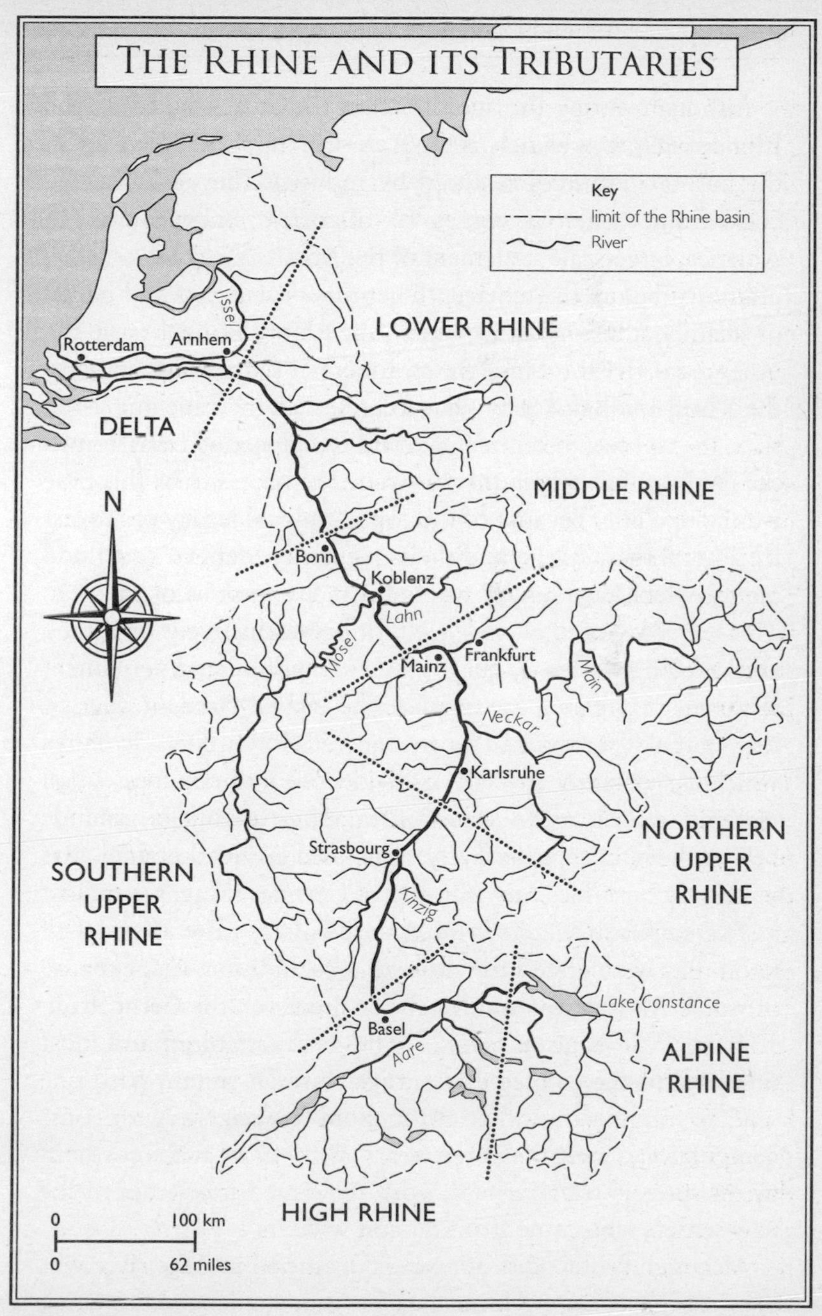

THE RHINE AND ITS TRIBUTARIES
Key
limit of the Rhine basin
River
LOWER RHINE
DELTA
MIDDLE RHINE
NORTHERN UPPER RHINE
SOUTHERN UPPER RHINE
ALPINE RHINE
HIGH RHINE
N
Rotterdam
Arnhem
Ijssel
Bonn
Koblenz
Lahn
Mosel
Mainz
Frankfurt
Main
Neckar
Karlsruhe
Strasbourg
Kinzig
Basel
Aare
Lake Constance
0
100 km
0
62 miles

Map 2

Although during the age of steam the area drained by the Rhine system was as rich as the American upper Ohio valley in the natural resources required by industry, they were hardly comparable when it comes to historical background. In America, large-scale settlement of the Mississippi river system – unknown before the nineteenth century – was part and parcel of steamboat history. In Germany, the Rhine, as much as all the other great rivers – Ems, Weser, Elbe and Oder – flowing into the North and Baltic Seas – had been settled for millennia. Even so in the eighteenth century hydraulic engineering transformed the land through which they flowed. The necessity by this time was compelling. Because of the topographical legacy of the last ice age, all these rivers, and their tributaries, defined vast flood plains, which were under water for many months of the year. This may have been all very well for harvesting reeds or fishing and catching wildfowl, but it made agriculture and permanent settlement extremely precarious. The typical German river, at least in the dry season, either meandered slowly between banks which it constantly eroded, or divided up into countless small streams, which, as they joined together again, left islands behind them. Local geography, as defined by either system, was inherently unstable. Any number of German villages were lost to rivers as their waters eroded their banks, while in times of flood they were often little more than islands in a vast expanse of water. All this was too much for Frederick the Great, from 1740 to 1786 King of Prussia – the best organized and most powerful of the eighteenth-century German states. With one scheme following another throughout his reign – with some interruption during the Seven Years' War (1756–63) – northern Germany was transformed, with reclaimed land opened for new settlers who came from far and wide.[5]

Although Prussia did not extend to the Rhine, the river was beset by the same problems as those that challenged Frederick the Great. Attempts to solve them went far back in history. As early as 1391, a channel was cut across a big loop of the river as it meandered south of Mainz; a cut of this kind was always a local solution, saving only a village or two from the ravages of the river, but even so the procedure was standard for centuries – there was no obvious alternative. In the late eighteenth century, Johann Gottfried Tulla, a young engineer from Karlsruhe, seeing what had already been achieved in north Germany – where Frederick the Great's strategy by Tulla's day had been widely followed outside Prussia – turned his attention to the Rhine. In 1806 the political climate became much more favourable: Napoleon, at the height of his power in Europe, dissolved the Holy Roman Empire of the German Nations, and in place of the countless small principalities that constituted it – particularly along the upper Rhine – established a small number of larger and much more powerful states. As a result, Tulla – as many others – became a citizen of the Grand Duchy of Baden, a state commanding the upper Rhine from Basel to Mainz, but with France on the other side of the river as far as Strasbourg. Fearing at first to be out of favour in the new Grand Duchy, he spent three years (1809–12) in Switzerland where he wrote *The Principles according to which future work on the Rhine should be conducted* – a massive treatise on practical hydraulics.

The essential principle was to carry out, on a much greater scale, what had been standard practice for centuries – which added up, in effect, to turning sections of the river into canals. Supported by Napoleon on the left bank of the Rhine, and the Grand Duke of Baden on the right bank, the way was open for Tulla to carry out his plans after he returned to Baden in 1812; a joint Magistracy of the Rhine had already been established in 1809. But then with the final defeat of Napoleon at the Battle of Waterloo in 1815, the whole project became unstuck. Tulla was none the less able to proceed with those of his plans relating to the Rhine north of Strasbourg – where it was entirely within the domain of the Grand Duchy – and continued his work until his death in 1828. It was only in 1841 that the Rhine Boundary Treaty allowed 'corrections' to proceed south of Strasbourg, and the complete operation – as contemplated in Tulla's treatise – was not finished until the 1870s, long after his death. In the middle years of the nineteenth century thousands of labourers worked with spades to dig the required cuts, consisting mainly of 18–24-metre wide channels. They were often joined by soldiers, who when necessary protected them from hostile local inhabitants. Horses supplemented human muscle, but help from steam-powered machinery only came in at the end of the 1860s. After every cut was made, the constant flow of the river made it both wider and deeper until after some five years, on average, the banks were secured by fascines consisting of bundles of brushwood tied together. As a result of this process the length of the Rhine between Basel and Worms was reduced from 220 to 170 miles, with any number of cuts and 150 miles of main dikes: at the same time 2,200 islands simply disappeared.[6]

How then did all this relate to the introduction of steamboats onto the Rhine – something which was never part of Tulla's plans? The first steamboat to be seen on the Rhine had sailed from London in 1816, and ended up in Frankfurt. With the *Concordia* a service opened up between Cologne and Mainz, carrying 18,000 passengers in its first year, 1827 – four years ahead of the first steamship on Tulla's upper Rhine.[7] The first lake steamer – on Lake Constance – also started service in 1827,[8] a precedent soon followed throughout Germany. The *Concordia* proved to have a very popular route, including the gorge with the Siebengebirge, famed for the Wagnerian scenery offered by the rocks of Drachenfels and Lorelei: by mid century, with several new boats serving the route, a million people took the trip.[9] This section of the Rhine, together with the Ruhr, a major tributary, was also becoming important for carrying coal from the local mines. For both passengers and freight it was fraught with hazard, above all at the Bingen Gap, where the reef had caused many shipwrecks when the water was low. Here Tulla's strategy had little to offer and dynamite was needed to solve the problem, by tripling the width of the gap.[10]

By this time, as propellers began to replace paddle wheels, the steamship had become the 'key to the changing shape of German rivers'.[11] Tulla's basic strategy still held good, but had to be carried out on a much larger scale – a process that still continues. At the same time the character of the traffic, particularly in freight, changed radically. The old 'uncorrected' river scene was dominated by timber floated downstream in the form of vast shallow rafts. With the steam age this traffic declined steadily: industry, and in particular shipbuilding, replaced wood with iron and steel, and such wood as was still needed was generally carried by rail. At the same time, the deeper, straighter river invited steamboats to bring cargo upstream, so that by 1907 this accounted for three quarters of the freight carried, a complete reversal of the proportions in the 1840s.[12] By this time some 85 per cent of the flood plain was lost, and as the river became deeper, so did the water table. At the same time the rate of flow increased, so that at the beginning of the twentieth century flood waters took only three days to cover the distance between Basel and Karlsruhe; by mid-century they only needed half this time, but there was a price to pay in the lower river when its banks could not contain the much more frequent floods – a phenomenon that can be observed all too often as the Rhine passes through the Netherlands to the North Sea. By this time the day of steam was long past, although the volume of traffic, often carried in rafts or lighters pushed upstream by a diesel tug, has not abated.

Although it is far beyond the scope of this chapter to look at all the inland waterways, worldwide, where steamboats entered service in the nineteenth century, even a general review is difficult – given the vast range of different geographical and economic cases. The extreme diversity of the great navigable rivers outside Europe and North America illustrates the point – as can be seen by looking at six of the best-known: the Amazon, Congo, Nile, Ganges, Mekong and Yangtze. As often as not they have, however, one thing in common: the first steamboats that plied them were made in Britain, to be shipped out in parts and assembled locally. Nowhere else in the world can ocean-going ships proceed inland so far as along the Amazon and its major tributaries. Iquitos, in Peru – more than 1,860 miles from the Amazon's Atlantic estuary – is regularly served by cruise liners, whose passengers sign up to enjoy, above all, the great rainforests, for which the river is well known. Navigation is not entirely straightforward, since the considerable seasonal variation in the level of the river makes the main channels difficult to follow. For many months of the year steamboat passengers must be content with endless stretches of water, with the rainforest little more than a distant horizon. Economically speaking, the greater part of the Amazon basin has little to offer either in the way of natural resources, or in prospects for commercial or industrial development – a fact that Brazil, which controls the greater part of the river system, often chooses to ignore. For some seventy odd years, starting with the second half of the nineteenth century, the fact that the rainforests were the world's only source of natural rubber assured the river – and particularly the city of Manaus, located at its confluence with a major tributary, the Rio Negro – unprecedented prosperity. Once South-East Asia was able to supply plantation rubber – only possible after seeds had been successfully smuggled out of Brazil in the late 1890s – Amazon rubber lost the greater part of its export market, and there was little to replace it. After their first appearance more than 150 years ago, the steamboats are still there. Watching them pass by are the members of countless Indian tribes, who for centuries have known how to make a living in the rainforest – often, moving their settlements in the face of main-stream Brazilians, anxious to exploit the land for agriculture and ranching, for which it is little suited.

The Congo, in a way, is an African version of the Amazon: a great river, with countless tributaries, following a long course through tropical rainforests. Unlike the Amazon, rapids close to the mouth of the river made access impossible for sea-going ships. The result was that ships to be found on the river, such as that described by Joseph Conrad in the passage cited at the head of this chapter, tended to be much smaller. In a sense they defined the standard for all the rivers of equatorial Africa, such as was familiar to film-goers in the 1950s when they saw Humphrey Bogart and Katharine Hepburn playing the leading roles in *The African Queen* – with a script adapted from a novel by C.S. Forester in which the action takes place during the First World War. Another critical difference from the Brazilian river scene was that the Congo was a European colony: the colonial power was the King of Belgium, who ruthlessly exploited the colony's African population and natural resources.

Although also in Africa, the Nile – one of the world's greatest rivers – has little in common with the Congo. If its two main sources, that of the White Nile in the three great lakes of equatorial Africa and that of the Blue Nile in the mountains of Ethiopia, are in areas of abundant, but seasonal, rainfall, its main course – particularly as it was known to a very long history – is through desert. Quite unlike the Amazon or the Congo, the economic importance of the Nile was always to be found in the use made of its flood-waters, as they rushed downstream in the late summer, for irrigation. The agriculture that this made possible led not only to very considerable populations settling along the banks of the river, and throughout its vast delta in lower Egypt, but also to the emergence of powerful autocratic states, such as those of the Pharaohs, whose monuments – notably the pyramids – still survive more or less intact. It hardly need be said that boats, some with sails – such as the stately feluccas – and others with oars, were always part of the river scene. Nor is it surprising that with the coming of the age of steam, Europeans, busy with exploiting the region's economy with the connivance of the local pashas, lost little time in introducing steamboats. Here, as with the Americans on the Mississippi, they were at least a generation ahead of the first railways.

The operation was not entirely straightforward. For one thing, the pronounced variations in seasonal flow made navigation difficult for much of the year. As on the Mississippi, boats with a shallow draught were part of the answer, but upriver from the first cataract, just above Aswan, sandbanks at the end of the season of low water were still a formidable obstacle: as the waters receded many a steamboat was left high and dry until the next floods came.

On the upper Nile, above Aswan, the river, flowing downstream from the Nubian deserts of Sudan, defined the route for the advance of British imperialism – for which the steamboat was indispensable. In this case geography insisted that railways were part of the operation, as is apparent from

Winston Churchill's description of his 1898 journey to join Kitchener's army in the Sudan in time for its planned confrontation with the Mahdi's Dervishes:

> We were transported by train to Assiout; thence by stern-wheeled steamers to Assouan. We led our horses round the cataract at Philae; re-embarked on other steamers at Shellal; voyaged four days to Wadi Halfa; and from there were proceeded 400 miles across the desert by the marvellous military railway whose completion had sealed the fate of the Dervish power.[13] In exactly a fortnight from leaving Cairo we arrived in the camp and the railway base of the army, where the waters of the Atbara flow into the mighty Nile.[14]

The railway referred to by Churchill, which cuts off the great Dongola bend of the Nile – so bypassing no less than three cataracts – follows a straight line across the desert. From its southern terminus the British army, to reach Omdurman – where the Mahdi's main force was waiting – had to advance some 200 miles overland on the left bank, having crossed the river at Atbara; every evening, for two weeks, the soldiers went down to the river, not just for water, but for supplies brought upstream – with an escort of Royal Navy gunboats – by 'stern-wheel steamers, drawing endless tows of sailing boats'.[15] With such support, some 20,000 soldiers – when they finally confronted the much larger forces of the Mahdi outside Omdurman – won a decisive victory. No one doubted that the supply lines, utterly dependent on steam-powered transport – by rail or river according to local circumstances – were essential. Once again the demands of military strategy were a key factor governing transport in the steam age.

India, in this respect, is the classic case. Its greatest river, the Ganges, cried out for steamboats – so much so that Robert Fulton, who from his base in New York had become the major local operator, was ready to repeat his success in India. For this purpose he found a business partner, Thomas Law, and in a letter of 16 April 1812 outlined his plans:

> I agree to make the Ganges enterprise a joint concern. You will please to send me a plan how you mean to proceed to secure a grant for 20 years and find funds to establish the first boat. This work is so honorable and important. It is so grand an Idea that Americans should establish steam vessels to work in India that it requires vigor, activity, exertion, industry, attention, and that no time should be lost. My Paragon beats everything on the globe ... this Day she came in from Albany 160 miles in 26 hours, wind ahead ... Keep the Ganges Secret.

Fulton died long before anything could come of his plans, and when steamboats did come to India, it was the British who built and operated them. Although as early as 1825 the East India Company was already contemplating a coastal steamer service, little came of its plans.[16] Even so, when steamers first appeared on Indian rivers in the late 1830s, they were still a generation ahead of railways. Among those involved was Robert Stephenson who – in spite of his fame (shared with his father) as a railway engineer – was also an outside contractor for shipyards. In the 1850s, not long before his death, he retained Stanton Croft and Co., solicitors in Newcastle, in an action to recover money owed by Charles Mitchell & Co., one of the largest Tyneside shipyards, for 'engines and machine parts' supplied for three Indian River Steamers.

As with Africa, the boats destined for India were shipped out in parts for local assembly. For a riverboat, under its own steam, an ocean voyage would be much too hazardous. It is, however, worth noting how in the days of sail – going back to well before the Christian era – boats sailed down the Ganges and continued across the Bay of Bengal to such distant destinations as Malacca.[17] (With steamboats in the nineteenth century the Amazon was the only long river where this was feasible.) As on the Mississippi, a steamboat built to operate on almost any Indian river had to have a very shallow draught. Although, as in Egypt, river steamers were introduced to serve British interests – both economic and strategic – local indigenous populations accepted them as readily as they did railways a generation later. What is more, the Ganges – including its many tributaries – was, just like the Nile, home to ancient civilizations with their own traditional craft on the river. Some lost out to the steamers – particularly when it came to long-distance travel – but many found a niche where they could continue to operate, often as ferries. This was still important, for India was pre-eminently a country of travel overland – as shown in all its variety in Rudyard Kipling's *Kim* – so the occasional ferry-crossing was the most that many ever experienced of travel by water. It also explains why, at the end of the day, rail travel – as described in Chapter 10 – was bound to win over river travel.

River steamboats met their greatest challenge on the Mekong river in South-East Asia. The Mekong is a long river with its source – close to those of the Yangtze and many other long rivers flowing into the western Pacific – in the mountains of south central China. From this point it flows through Laos (where it is for some distance that country's frontier with Thailand) and Cambodia, finally to reach the South China Sea via a vast delta in southern Vietnam. In a number of stages – starting with the seizure of the Vietnamese seaport of Saigon in 1859 – France, in the second half of the nineteenth century, appropriated Vietnam, Cambodia and Laos, and in 1887 incorporated them into a single colonial administration, known as the Indo-Chinese Union. In geographical terms the Mekong was the only unifying feature of the Union: the populations along its bank were – and still are – extremely diverse. What is more, the various principalities to which they belonged before the French took over had a long history of war, so that, for instance, by the nineteenth century the Kingdom of Cambodia – as a result of a succession of defeats – was much smaller than the medieval Khmer Empire which had preceded it. Indeed the Khmers had for a long time ruled the Mekong delta, which now reaches just short of the present Cambodian capital of Phnom Penh.

To the French, it was apparent, at a very early stage, that if they were to make anything of the Mekong, they should introduce steamboats, but the geography of the river, with a succession of cataracts, did not favour any such project. After he had returned from his travels up the Mekong, Louis de Carné, an intrepid explorer, and a member of the Mekong Exploration Commission – which in the late 1860s, conducted a remarkable survey of the river along almost its whole course downstream from the Chinese province of Yunnan to the South China Sea – related that:

> the truth at last began to force itself on the most sanguine among us. Steamers can never ply the Mekong as they do the Amazon or the Mississippi; and Saigon can never be united to the western provinces of China by this immense riverway ... However magnificent [the river] seems only to be an incomplete masterpiece.[18]

Given its strategic importance, the French persisted in their attempts to send steamboats up the Mekong. In doing so they consistently understated the obstacles facing them, even though they accepted that in the Mekong's 'gigantic landscape everything exuded unimaginable power and assumed crushing proportions'.[19] The most formidable obstacle was the Khon Falls at the point upstream where the Mekong first becomes the frontier between Laos and Cambodia. The falls were just one of the chutes to be found among the cataracts which, according to their local name, *Siphandon*, divide up the river into 'The Four Thousand Islands'. The French believed that a route navigable by steamboats could be set up by making use of channels known to the local population. Four times in the period 1892–4

> diminutive steamships attempted the impossible. With engines roaring and boilers near bursting, with hundreds of men hauling from rocks with ropes and others pushing from the decks with pikes, the steamers addressed the Falls ... Although the ships were of little more than one metre's draft and fourteen metres long, they all either grounded or were fouled by floating timbers.[20]

In 1893 two steamboats, the *Massie* and the *Lagrandière* – their parts constructed in France but assembled in Saigon – proceeded upstream to the foot of the Khon Falls. There they were disassembled, and the separate parts loaded on to flat trucks on a short length of railway on Khon Island. In the absence of a locomotive manpower was the only means of pulling the trucks over the steep gradients leading to the upstream shore of the island. When there proved to be too few rails to complete the track, those behind the trains were pulled up and laid in front of them, so enabling the two boats to reach the Upper Mekong, to be reassembled. Both then entered service on the Upper Mekong, but a high price was paid for this *prouesse d'acrobatie nautique*.[21] It cost not only a lot of money, but also the lives of hundreds of Vietnamese labourers impressed into service on Khon Island at the height of the malaria season.

Both boats were ill-fated. Captain Simon of the *Lagrandière* – suffering from the all-too-prevalent *folie de grandeur* – took his boat upstream through Laos with the intention of reaching China. While still in Laos the Tang-ho cataract – upstream from the capital, Luang Prabang – proved to be as formidable an obstacle as the Khon Falls. Once Simon accepted defeat, he found his boat marooned as a result of the river level falling and so had to wait nearly a year before new flood waters allowed him to return downstream. The *Lagrandière* continued in service until it was lost with all hands after hitting a rock just south of Luang Prabang. The fate of the *Massie* is unknown.

In spite of this inauspicious start, the whole enterprise did point the way to establishing steamboat services up the Mekong at least as far as Luang Prabang. In 1897, a railway, with a single wood-burning locomotive – assembled at the bottom of the Khon Falls from parts sent out from France – opened to transport both passengers and goods along the same route as that used by the *Massie* and *Lagrandière*. At Khon village, on the Upper Mekong, a new steam launch, the *Garcerie*, waited to take them upstream. The railway was extended in the 1920s to Don Deth, an island immediately upstream from Khon, where steamers had fewer problems from low water levels. To facilitate the transhipment of goods steam-powered cranes were installed at both terminals of the extended railway, but even with this facility its days were numbered. The last train ran in 1940; occasional stretches of track are still to be found on the two islands as is also the bridge between them. By the end of the Pacific War in 1945 the whole of Indo-China was under Japanese control, and even with the defeat of Japan France never recovered its empire in South-East Asia. The fifty odd years of steamboats on the Mekong could never be counted among its successes.

At least in the final years of the Age of Steam, steamboats on the Yangtze and its tributaries could compete, worldwide, with those of any other river system. In China itself the river has always been unrivalled, with hundreds of millions of people living in its catchment area. It is navigable by ocean-going steamships at least as far as Wuhan, which is some 620 miles inland, and upstream beyond Wuhan is open to river steamers over an even greater distance. In the seventy odd years from 1842, when the Treaty of Nanking recognized the first five Treaty Ports – including Shanghai, which is close to the mouth of the Yangtze – ten of the fifty were located along the river, and of these even Chongjing, the furthest inland, was served by steamers. Seeing that the Chinese Qing dynasty – which survived until 1911 – was always compelled to respect the privileges claimed by the nations which had established the various treaty ports, it could do little to impede steam navigation. As Chapter 11 shows, this is a quite different case to that of the railways, whose development was long held up by traditional aversion to foreign technology. Until 1890 the imperial government did succeed in restricting steam traffic on the Upper Yangtze – so that steamboats did not reach Chongjing until that year – but resisting steam on the main rivers was always a losing battle. The economic advantages were overwhelming: Shanghai was the only coastal port that handled more traffic than Wuhan and Jiujiang, the two main ports on the Yangtze open to ocean-going vessels. In agriculture, grain could not reach its markets without the Yangtze steamboats, and industry was equally dependent upon them for the supply of coal and raw materials.[22]

The Yangtze river scene was always very mixed: in sheer numbers small traditional craft – including fishing boats – always dominated, and in the twentieth century heavy goods traffic inevitably lost out to railways. Throughout the era of the treaty ports, foreign powers dictated the character of steam traffic on the Chinese rivers; this was particularly true of the Yangtze, the most important and accessible of all of them.

7

THE RAILWAYS' FIRST FORTY YEARS

In the early stages, railways were planned, financed and finally constructed to satisfy specific economic demands, adopting the rationale of the Liverpool & Manchester Railway as described at the beginning of Chapter 8. On this basis constructing a railway was essentially a niche operation, but the business success of the earliest railways, both in Britain and abroad, meant that by the end of the 1830s any number of entrepreneurs were looking for new niches. By this time, however, some of the biggest projects were already nearing completion. Already, in 1838, Euston had opened as the first London terminus of a main line – that of the London and North-Western Railway. If the Great Western Railway, with its main line from London to Bristol – which was designed to give Bristol the same advantages as a gateway to the Atlantic as Liverpool had gained, in 1830, from its rail link to Manchester – is now better known, the reason is to be found in the character and vision of the man who was its first chief engineer.

Isambard Kingdom Brunel was the workaholic son of a refugee from revolutionary France, Marc Brunel, who, once settled in England in 1799, established a very considerable reputation as an engineer ready to take up any challenge – such as building the first ever tunnel under the River Thames. The son was soon to exceed his father's achievements, notably with his work for the Great Western. His vision was of a railway operating with maximum efficiency and capacity, to be achieved first by a route avoiding any but the most gradual curves and gradients and second by locomotives and rolling stock of a much wider gauge than the standard established by the Stephensons, who, in the 1830s, still dominated every aspect of railway engineering.

History has fully justified Brunel's choice of a route north of the Marlborough downs, with its own new town of Swindon for the railway's workshops, but – although he was never to know this – his wide 7-foot gauge in the end failed to hold out against the Stephensons' standard of 4 foot 8½ inches. Brunel's engineering on the new line was remarkable, not only for such achievements as the Wharncliffe Viaduct, just outside London, the bridge over the Thames at Maidenhead[1] and the Box Tunnel – nearly 2 miles long in west Wiltshire – but also for the unprecedented scale of his operations, with thousands of labourers employed on construction. In fact Brunel was a better builder than operator; he was the despair of his directors because of his handling of accounts and of his subordinates because of his incapacity to delegate, and in spite of his being the first to use the electric telegraph for railway operations he was blind to its full potential – as demonstrated by W.F. Cooke – for ensuring their safety and efficiency. On the other hand, as Chapter 12 will show, his vision and achievements went far beyond railways. He also had one particular success: it was on his Great Western Railway that a reigning British sovereign travelled for the first time by train. On 13 June 1842, Queen Victoria travelled by train from Slough to Paddington, a journey that took twenty-five minutes – a time much exceeded by that taken up by the ceremonies at either end. Brunel was believed by many to have been the engine driver: not so – while he was certainly on the footplate of the locomotive named 'Phlegethlon', the actual driver was his chief engineer, Daniel Gooch.[2]The last two years, 1839–40, before the Great Western Railway opened on 30 June 1841 were the time of the first railway 'mania', in which investors, inspired by such giants as Brunel, invested a sum equal to nearly 2 per cent of the national income in railways. This, however, was little compared to the 7 per cent of 1847, the peak year of the second railway mania, in which more than 6,000 miles[3] of line were built and more than a quarter of a million men – some 4 per cent of the male workforce – employed on construction.[4] The result of such high levels of investment was to transform a disjointed complex of mostly short separate lines – each with its own distinctive niche – which was the position after the first mania, into something close to a comprehensive network covering the whole of Britain. The result in fact was considerable overkill, as had been foreseen, and much regretted by George Stephenson and many others among the railway pioneers. Brunel went so far as to say, 'I wish I could suggest a plan that would greatly diminish the number of projects; it would suit my interests and those of my clients perfectly if all railways were stopped for several years to come.'[5]

The quotation at the head of Chapter 8, which comes from Dickens' *Dombey and Son*, shows how drastic was the impact of this process during the 1840s. Its social implications were to be seen in vastly increased urban populations, largely the result of migration of unskilled labourers from the countryside but also from Ireland in the decade of the great potato famine. At the same time, however, education was making great gains among the working classes, which were gradually being transformed into a literate population. Infectious disease, notably cholera, was rampant in the crowded inner cities, and it was only later in the century that it was brought under control.

The economic consequences were just as significant. While it can be argued that railway construction diverted investment from other targets, its vast demand for goods and labour must have accounted for considerable economic growth. The supply of tens of thousands of standard lengths of wrought-iron for rails, millions of bricks for bridges, tunnels and cuttings, and tons of coal to be consumed by locomotives and railway workshops, was only possible with vast industrial expansion – while, at the same time, only the new railways could provide the necessary transport. Railways also made possible new forms of traffic, such as the transport of perishable goods – meat, fish and vegetables – to the great city markets, such as London's Smithfield, Billingsgate and Covent Garden. From 1838 special trains carried a breathtaking volume of post for next-day delivery, a service only finally suspended in 2003. (Forty years earlier, in 1963, it had provided the occasion for the great train robbery in which some £8,000,000 was stolen from the Post Office.) The travelling post office became absolutely indispensable following Rowland Hill's great reform of 1840 – the introduction of the penny post paid in advance by an adhesive stamp. He had conceived of this as a particular benefit to the new urban literate classes, anxious to remain in touch with their home villages at an affordable price.

Finally, throughout the nineteenth century railways were a great stimulus to horse-drawn traffic. The transport firm of Pickford's, which in the eighteenth century had transported goods both by road and canal, flourished as never before in the nineteenth century by providing short-haul horse-drawn transport to railway goods stations.[6] Parallel to this, hansom cabs, in their hundreds, brought passengers to the stations, while the horse-drawn omnibus provided public transport in the transformed cities of the new railway age. Is it then so surprising that the number of horses actually increased substantially in the nineteenth century?

Extending the survey of new railways to continental Europe and North America involves a consideration of so wide a range of different political, economic, geographical and social situations, that few generalizations are possible. The way these factors combined, in the case of Britain, is important if only because Britain came first. If the earliest continental railways, say those founded before 1840, originated as independent niche operations – as had been the case in Britain – it was soon clear that quite distinctive local factors would determine their future development. The governments of European states were much more pro-active in using the power of the state whether in encouraging and controlling railway construction or in frustrating it. The case of Belgium is exemplary: the industrial potential of a newly independent state, densely populated and with abundant resources of coal and iron, was a very powerful stimulus to railway construction – both for passengers and goods – strongly encouraged by its first king and his ministers. The result was the world's densest national railway network, which in the twentieth century became as much a burden as a blessing to the Belgian economy. There was also a troublesome political legacy: for geographical reasons, based on the location of natural resources, new industry was concentrated in such cities as Mons, Charleroi, Namur and Liège, which were all French-speaking, so that the Dutch-speaking Flemings – a majority of the population – shared relatively little in the prosperity brought by the railways. This at least was their perception, but in fact the first line to be constructed – which linked Brussels, the capital, to Antwerp, the largest port – was entirely in Flanders. (Economically speaking, the tables were turned in the course of the twentieth century, as the nineteenth-century economy of French-speaking Belgium, based on coal and iron, declined while new light industry, largely dependent upon road transport, developed strongly in the Dutch-speaking part of the country.)

The railway history of Italy provides another illustration of power play. Until the 1860s Italy was a heterogeneous collection of political units, of which the most progressive was undoubtedly the northern Kingdom of Piedmont, with either the papal states at the centre of the peninsula or the utterly decadent Kingdom of Naples at its southern extreme the most reactionary. Piedmont was always in favour of railways, not just for economic reasons, but also for the contribution they could make to the unification of Italy – a cause that dominated the kingdom's politics in the 1850s under Count Cavour, its prime minister. The papal curia of the mid nineteenth century, true to its unenlightened character, did not favour railways for the considerable territories it governed. When Pope Gregory XVI died in 1847, this led to a joke told between the many Romans who longed for a railway. Pope Gregory, walking up the dusty road leading to heaven, met Saint Peter and asked him if there was still a long way to go. 'Much farther yet,' was the answer, 'but if only you had built a railway you would be in paradise by now.' Even when the cause of unification triumphed in the 1860s, the entire railway system of the Papal States consisted of two short lines, each less than 60 miles, linking Rome to Civitavecchia on one side and Frascati on the other. With no link whatever to the extensive network in northern Italy – much of which belonged to the Austrian Empire until the end of the 1850s – Rome had little benefit from the prosperity that railways had brought to the north.

The Austrian case is also interesting: as opposed to Italy, where unification was the order of the day, Vienna was concerned to hold together a large empire comprising many different ethnic groups – Italians, Czechs, Hungarians, Poles to name but a few – all open to the tide of nationalism that was sweeping Europe. Railways had a clear advantage as an instrument of government – particularly for their capacity for the rapid movement of troops. In 1850, because Austria scored in its stand-off with Prussia for precisely this reason, the terms for settling the dispute between the two countries laid down by the Peace of Olmütz, were particularly favourable to it.

The point was well taken by the Emperor Franz Josef (1845–1917) who, although at heart hardly more enlightened than the Papal Curia, well realized the strategic potential of railways. (To please the Emperor a special line was constructed to link his palace at Schönbrunn with Vienna and for the opening ceremony he travelled to the new station at Schönbrunn in a special train, the only occasion on which the line was ever used.) While, topographically, the rail link to Prussia, which was north of Vienna, had been comparatively straightforward, any railway built south or west of Vienna – as the war against Piedmont in 1859 would show, they were directions of equal importance strategically – would encounter formidable natural obstacles. Of these the first was the Semmering Pass across the eastern Alps, less than 60 miles south-west of Vienna. The engineer in charge, Karl Ritter van Ghega, planned a route with 1 in 40 gradients, but first consulted George Stephenson as to whether a locomotive could achieve such a steep climb. Stephenson was negative, but Ritter van Ghega – having looked at routes already built in America – refused to take 'no' for an answer. Instead, in 1851, he arranged for an Austrian retake of the Rainhill trials (won by Stephenson's 'Rocket' in 1829) with a prize of 20,000 imperial ducats for the best-performing locomotive on the Semmering Pass. Four locomotives were entered, one from Bavaria, one from Belgium and two from Austria – with the prize going to the first of these, aptly named 'Bavaria'. (The fact that all four locomotives were well up to the assigned task shows how much had been achieved in the first quarter-century of the railway age.) The 'Bavaria' then opened the line, the first across the Alps, in 1854: in 1857 this was extended to Trieste, to complete the first railway route linking – if somewhat circuitously – the Adriatic with the North Sea.

In Prussia Kaiser Wilhelm I also favoured the construction of railways for their military advantages, but in this case his state, whatever its ambitions, was but one of the countless independent principalities that constituted Germany before 1870. As early as 1849 Prussia had sent three battalions of soldiers by train to Dresden, to help the King of Saxony suppress a popular uprising. A year later, in 1850, faced by the threat of war with Austria, Prussia's attempt to use railways to concentrate troops in the south of the country resulted in chaos, leaving troops scattered around the country, often without access to food, water or latrines.[7] At the end of the 1850s, General von Moltke, recently appointed Chief of the General Staff and convinced of the strategic importance of railways, tried hard to bring order out of chaos. This, in his view, became essential in 1859 when the

Kingdom of Piedmont, in Italy, with the support of France, went to war in order to free Lombardy and Veneto from Austrian rule.[8] Prussia responded by mobilizing its armies in order to support Austria: since this could involve war with France, von Moltke planned to move a quarter of a million Prussian troops by train westwards to the Rhine and the Main. The Kaiser dithered, the bureaucracy dragged its feet, and by the time von Moltke was free to move Austria had been decisively defeated at Magenta in northern Italy. For von Moltke it became even more urgent not to lose time, but then Berlin insisted that he should first gain support from the rest of the German Confederation. This was reasonable enough seeing that the destination of the Prussian troops was far outside Prussia, but by the time this hurdle was crossed, Austria and France were on the point of agreeing an armistice. What is more, the rolling stock required for the original mobilization had been scattered across Prussia to meet the normal demands of its railways.

At the heart of the problem, from Prussia's point of view, was the fact that the German railway system was the result of any number of uncoordinated projects, each of which had its own rationale, depending on local political and economic factors. Although, as a result of the lessons learnt from von Moltke's mobilization, there was some strategic coordination during the 1860s, the legacy for the country as a whole, after unification in 1870, was to be found in relatively backward main-line and long-distance services. Even so, von Moltke's influence then made certain that the network as a whole developed more rapidly than that of France. At the same time, however, priority for capital expenditure was given to improving the navigation of the major rivers – as described in Chapter 6.

France was a quite different case: the country, in spite of considerable political and social unrest, had long been united –

although its frontiers had changed often enough – and both popular and commercial demand for railways had been strong from the very beginning of the railway era. In part this was inspired by France's fear of its industrial economy lagging behind Britain's, but in any case the country – in contrast to Austria and Italy – was well suited, geographically, for laying down railways. This is clear from the map of European railways in 1850 on page 196, which shows how the only integrated network was in the gently rolling landscape of northern France, with links to the well-developed Belgian network, and on to Germany and beyond.

As with the Stockton & Darlington Railway in England, the first French line built for and operated by steam locomotives – linking Lyon and St Etienne – was intended for the transport of coal. Marc Seguin, the engineer who built it, had always contemplated the *ligne de premier ordre* for rapid long-distance passenger trains. Inevitably, given the powerful French tradition of a centralized government based on Paris, the system envisaged by Seguin depended upon the bureaucracy taking the initiative. In 1842 this took shape in a law that provided for government to provide the infrastructure, consisting of the permanent way, including all bridges and tunnels, leaving the superstructure, consisting of rails, ballast, stations, locomotives, rolling stock and working capital, to private enterprise. Even here nothing was left to chance, since the lines were to radiate from Paris in six segments – like slices of a cake – with the divisions between them prescribed in the legislation. Each segment would have its own mainline station in Paris. It was then open to the private sector of the economy to organize the companies to operate the six segments – with the considerable advantage that the state, by taking responsibility for the infrastructure, provided them with a massive subsidy from public funds.

This bureaucratic arrangement, although reflecting the popular conception of France as l'Hexagon, quite failed to take into account its demography: the French population, which is extremely skew, is concentrated along the major rivers – particularly the Seine and the Rhône – the Côte d'Azur along the Mediterranean between Marseille and Italy, the coasts of Normandy and above all along a wide, heavily industrialized strip of land, containing such cities as Lille, Thionville, Metz, Strasbourg and Mulhouse, extending from the English Channel to Switzerland and bounded on its eastern side by the frontiers with Belgium and Germany.

Predictably investment interest concentrated on railways for the segments offering the best economic prospects, as determined largely by such demographic factors. The result was that the Chemin de Fer de l'Ouest, whose terminus at the Gare St Lazare opened in 1837,[9] and the Chemin de Fer du Nord, whose terminus at the Gare du Nord opened in 1846, led the field. While the former, with its lines to Rouen, le Havre and Dieppe, benefited from operating profitable services to upmarket Paris suburbs such as St Cloud and St Germain-en-Laye, the latter, with its lines to Calais, Boulogne and Valenciennes – the gateway to Belgium – was a much more complete system, and the only one with international connections. Somewhat oddly the first Paris terminus to open was the Gare d'Orléans, from which, as its name suggests, trains ran, more or less due south from Paris, to Orléans on the River Loire – a city which by the nineteenth century had lost much of its historical importance.

In one direction, that south-west of Paris, for which there were too few investors for an operating company, the state had to take over with its own Chemin de Fer de l'État. From 1855 this ran from the Gare de Montparnasse, which it shared with Chemin de Fer de l'Ouest. (The station is strictly known as Maine-Montparnasse to distinguish the lines served by the two companies; the station complex now includes the Tour de Montparnasse, the tallest building in Paris.)

A much more problematic line was the PLM – for Paris, Lyon, Marseille. In principle a railway linking the three largest cities in France, running south from Lyon down the valley of the Rhône, and with the prospect – soon realized – of extending along the Mediterranean coast to Nice and beyond, should have been a profitable venture. The distances, however, were long, and the terrain, particularly around Lyon, not at all easy. Even so, after a false start with a line to Dijon, the PLM was constructed, but it was only complete after a long tunnel under Lyon – now rivalled by an equally long tunnel for the Autoroute du Soleil – opened for traffic in 1855. (The Gare de Lyon, opened in 1853, is the most imposing of the Paris termini, famed for its station restaurant.)

The last of the six main lines to be completed served the east of France from the Gare de l'Est, opened in 1855. Given, on one side, the economic, political and strategic importance of eastern France – with the Rhine as its frontier with Germany – and on the other a countryside relatively favourable for railway construction, it is perhaps remarkable that it took so long to lay down the Chemin de Fer de l'Est. The two eastern provinces of Alsace and Lorraine were lost to France as a result of its defeat in the Franco-Prussian war of 1870. Germany, having once incorporated them, greatly increased the investment in their railways, leaving as a legacy a magnificent station at Metz, the capital of Lorraine.

Railroads in America

Railway development in America was from the beginning the result of the divergent economic interests of the separate states of the Union. Although each one of the thirteen founding states had its own port on the Atlantic seaboard, the advantage of making it, in any one instance, the hub of a railway network, depended on geography. The key distinction was between harbours such as New York and Philadelphia, which were the gateway to long navigable rivers already well-served by steamboats at the beginning of the railway age, and others, such as Boston, Baltimore and Charleston, which lacked this advantage. Such cities, then, had every incentive to support the construction of railways, to provide transport for their economic hinterland. On the other side of the line, New York – based on the island of Manhattan at the mouth of the Hudson River (which was navigable at least as far as Albany, some 160 miles upstream) – had little immediate incentive to invest in railways.

Baltimore, and the Baltimore & Ohio Railroad (B&ORR) – already mentioned in Chapter 5 in connection with their rivalry with the Chesapeake & Ohio Canal – provide the right background for studying how and why American railways developed in a way quite different from that of any European railway. Inspired by the success of the early British railways, the original directors of the B&ORR, having started off by adapting British technology (including the importation of British locomotives), soon found that principles established by the Stephensons – although often accorded the status of holy writ in Britain – were unnecessarily restrictive in the quite different topography of the hinterland of Baltimore.

First, and above all, was the matter of distance. That separating Baltimore from the Ohio river, the intended destination of the B&ORR, was hundreds of miles greater than the length of any railway proposed in Britain. When the B&ORR reached the Potomac river at Point of Rocks, Maryland – completing the essential first stage on the way to the Ohio –

the total length of its track already exceeded the combined mileage of the Stockton & Darlington and the Liverpool & Manchester railways in Britain.

By this time the directors of the B&ORR were laying the rails on wooden ties – in preference to the Stephensons' stone blocks – and equipping rolling stock with wheels with flanges on the inner edge fixed to rotating axles – both innovations eventually becoming standard worldwide. It was realized, also, that the British insistence on slight gradients and curves – almost an obsession with Brunel – would involve prohibitive costs in constructing any line across the Appalachians. The alternative was to build much more powerful locomotives: the local design, adopted by the B&ORR, had its weight evenly distributed over at least four driving wheels, with a swivelling front bogy to enable it to deal with the sharpest curves. The Stephensons' locomotives, which – being prone to derail – were only suited to level lines, plainly had no future in America. The fact that they were much more economical with fuel counted for little: in contrast to Europe, the first American railways – constructed at a time when coal-mining was little developed – followed the example of American steamboats and burnt wood, which was just as abundant along their lines as it was along the rivers. Later, when they did eventually switch to coal, this had already proved to be as abundant in the regions where they operated as it was in Britain: in the American case, as earlier in Britain, a number of important lines, such as the Lehigh Valley Railroad in Pennsylvania, were constructed specifically to serve coal mines.

Boston, in the very early days of the 1830s, was in the forefront of railway investment: by 1835 it was already a hub, with three different railways, each with its own terminus. Of these one, the Boston & Worcester Railroad, after linking up with the Western Railroad of Massachusetts, reached the Hudson River at Albany after crossing the Appalachians with what, in its day, was the highest railway in the world. Although, after its opening in 1841, this line operated at a profit, it was in the end a failure. This was in part the result of the intransigence of New York state, of which Albany was the capital. New York was against the line for two reasons: first, any extension to the west would not only threaten the profitability of the Erie Canal, but would have to contend with the 'fallacy that steam could run up hill cheaper than water could run down', and second, in the American export market any east-bound traffic to Boston would be lost to the port of New York. If there were to be railroads in the state, and they were soon constructed, they must serve its interests and not those of Massachusetts. Even so, the first rail link between Albany and Buffalo, formed by the New York Central Railroad, only opened in 1853.

In the southern states, Charleston, in South Carolina, in its attempt to develop its hinterland by constructing a railway, encountered much the same difficulties as Boston had. The best route inland needed to pass through Georgia, which was anxious to protect its steamboat traffic on the Savannah river. South Carolina was left to build its railways over the Great Smoky Mountains, which was achieved with a deviant 5-foot gauge line whose maximum gradient of 4.8 per cent was the highest in the US. Although this was within the powers of the newest locomotives, the inland terminal at Hamburg, on the Savannah river, hardly shared in the lucrative transport of cotton from the plantations that stretched west of the river as far as the Mississippi. Georgia, on the other side of the river, much preferred to see cotton – the mainspring of the Southern economy – shipped by steamboat downstream to its own seaport at Savannah. Then, when it saw the advantages of rail for going further inland, its own line, the Western and Atlantic Railroad, was extended north into Tennessee to a new terminus at Chattanooga – made famous by the 'Chattanooga Choo-Choo' – on the Tennessee river, a navigable tributary of the Ohio. In Georgia this railway started at a town appropriately called Terminus where its link with the Central Railroad of Georgia meant that the whole of the upper Mississippi river system had, from 1845 onwards, its first outlet to the sea in the southern states. When Terminus became much the most important railway hub in Georgia, the name of the city was considered too prosaic. It was changed therefore to Atlanta – recognizing the contribution of the Western and Atlantic Railroad to its prosperity. Atlanta became not only the largest railway town in the US, but also the capital of Georgia, and, in the second half of the twentieth century, the world's largest airport. In the nineteenth century, however, Savannah, even with its new railway links to the heart of the US, never attained the success of Boston, New York, Philadelphia and Baltimore.

By the end of the 1850s, of the six railways crossing the Appalachians only one was in the south. Of the remaining five four[10] were in the US and one in Canada,[11] a situation reflecting the secular decline of the southern economy in relation to that of the northern states. This was reflected in a length of rail in the north several times that in the south – a factor that proved to be decisive in the Civil War.

Before the war there had also been considerable railway construction west of the mountains, but – with one or two minor exceptions – east of the Mississippi. (The first railway bridge across the river was only built, at Rock Island, Illinois, in 1857.) When it came to freight, the necessity for such development was somewhat doubtful in the light of the steamboat traffic on the Mississippi and its major tributaries, as described in Chapter 5. On the other hand railways had already crossed the Appalachians to points on the Ohio where river cargoes could be transhipped for carriage to the east coast states and the Atlantic ports. This traffic was lost to the canals, rather than to the river, but it could only be a question of time before the railways crossed the Ohio and went on to the Mississippi. The B&ORR was the first to reach this goal, at East St Louis, Illinois, in 1857; in the next four years both the New York Central and the Pennsylvania Railroad reached Chicago, where the Rock Island Railway – with the first ever bridge across the Mississippi opened in 1857 – continued on to the great river itself. The 1850s, therefore, witnessed the start of the process by which Chicago would become America's most important railway centre.

Here a major part was played by the Illinois Central Railroad. The strategic location of a state that was bounded in its north-east corner by Lake Michigan, to the west by the Mississippi, to the south by the river's most important tributary, the Ohio, and to the east by the Wabash, a main tributary of the Ohio, was indisputable, and its citizens – mostly recent immigrants like Abraham Lincoln – were waiting to exploit it, which they did with considerable success. A railway system – designed mainly for carrying passengers rather than competing with the rivers for freight – that linked the south of the state at the confluence of the Ohio and the Mississippi with Chicago and Rock Island in the north, was seen as essential to economic development.

A novel solution was found for the problem of raising sufficient funds. If the Federal government in Washington, which – as in most of the new mid-western states – was the largest landowner, could allow the proposed railway to sell standard-size plots along its lines, then the money raised would pay for its construction; even better, those who bought and developed the plots would themselves generate new traffic, both passenger and freight. This was the origin of the land-grant system which, after the end of the Civil War, would pay for the construction of new lines to the west coast. As to Illinois, the scheme, although in principle acceptable in Washington, could only get congressional approval if a southern state was offered something comparable. The result was the Mobile & Ohio Railroad, linking the port of Mobile on Alabama's gulf-coast, with a point on the Ohio, opposite Cairo, Illinois, at the confluence of that river with the Mississippi. Although this established the first north-south rail link between the Great Lakes and the Gulf Coast, the line's success was limited. In part this was the result of the Civil War, but even before the war other rail links between the Ohio and the South, such as that of the Louisville & Nashville Railroad, profited more from trading cotton from the South and agricultural produce from the North – the essential economic rationale of all such links. This is the end of the story of the first generation of American railways, which ended with the Civil War, and which, while establishing a comprehensive network east of the Mississippi – particularly in the northern states – hardly extended beyond it.

Finally, the war itself deserves more than a footnote in American railway history. At the beginning of the war, in 1861, it was clear to the Union generals that when it came to rail transport they had an overwhelming logistical advantage. In 1860 the North had 21,978 miles of railway, compared to 9,010 miles in the South. In the same year only nineteen out of 470 locomotives were built in the South, where the railways, working with several different gauges, were much more disjointed. Then, as the war went on, both track and rolling stock had to be cannibalized by mainly unskilled labour to maintain a skeleton service.[12]

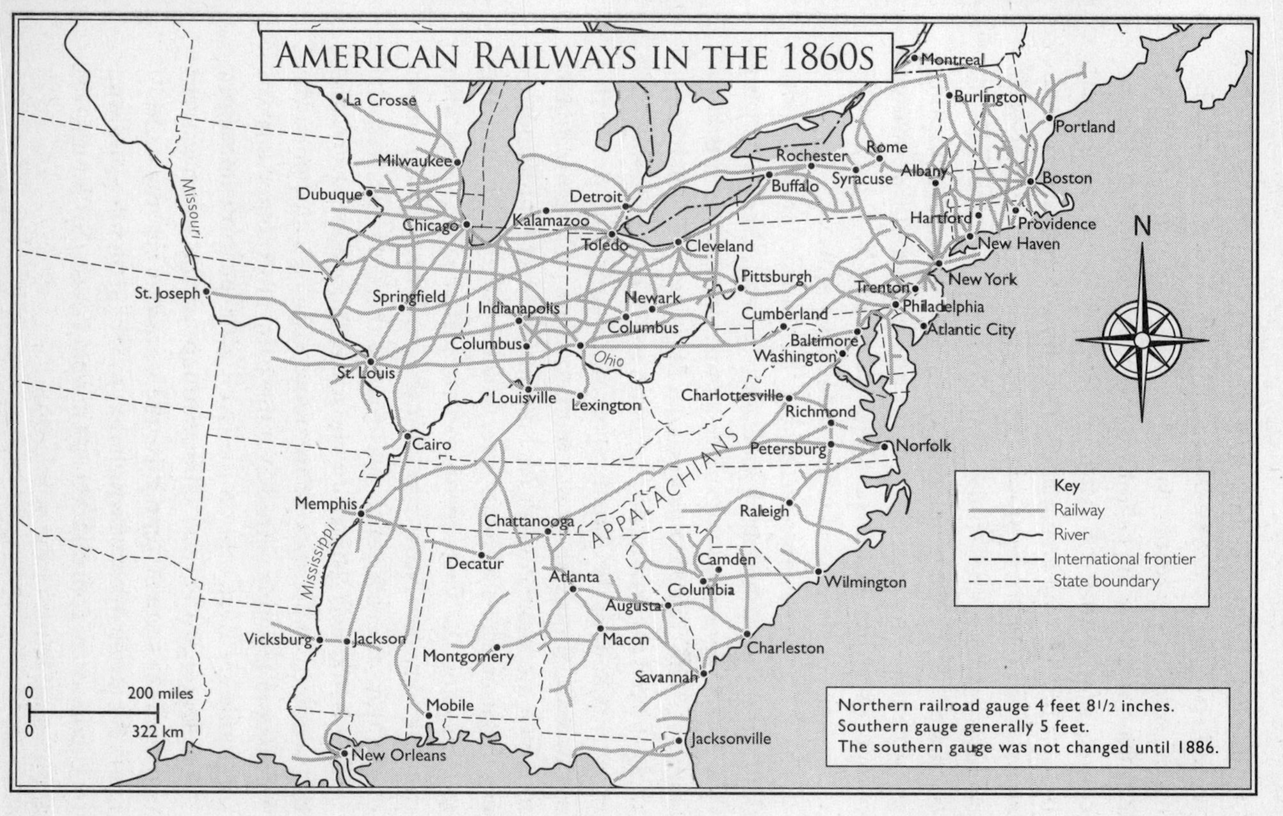
AMERICAN RAILWAYS IN THE 1860S
Montreal
Burlington
Portland
La Crosse
Milwaukee
Rochester
Rome
Albany
Boston
Buffalo
Syracuse
Missouri
Dubuque
Detroit
Hartford
Providence
Chicago
Kalamazoo
Toledo
Cleveland
New Haven
N
New York
Pittsburgh
St. Joseph
Springfield
Newark
Trenton
Indianapolis
Philadelphia
Cumberland
Columbus
Atlantic City
Baltimore
Columbus
Ohio
Washington
St. Louis
Charlottesville
Louisville
Lexington
Richmond
Cairo
Petersburg
Norfolk
APPALACHIANS
Key
Memphis
Railway
Raleigh
Chattanooga
River
Decatur
Camden
International frontier
Mississippi
Atlanta
Wilmington
Columbia
State boundary
Augusta
Vicksburg
Jackson
Macon
Charleston
Montgomery
Savannah
0
200 miles
0
322 km
Mobile
Jacksonville
New Orleans
Northern railroad gauge 4 feet 8 1/2 inches.
Southern gauge generally 5 feet.
The southern gauge was not changed until 1886.

Map 3

Although it was a Confederate general, Joseph E. Johnson, who first used railways for tactical purposes, by moving 6,000 soldiers up to the front just before the first Battle of Bull Run in 1861, the Union soon got the message, and in the course of the war, US Military Railroads, a new special branch of the War Department, supervised the transformation of the northern lines into 'an integrated, efficient, mostly double-tracked and fully standardized network'.[13] Its command of civil engineering was also remarkable, making possible the construction, in forty hours, of a 400-foot-long bridge over the Potomac Creek to bring the railway to Gettysburg in support of General Meade's army.

The most remarkable incident in the railway war was the Andrews raid, named after its leader, James J. Andrews, who conceived the underlying plan almost entirely on his own initiative. The concept was born out of the way in which geography forced the war to be fought on two fronts, separated by the great chain of the Appalachian Mountains. The war in the east was fought mainly in the southern state of Virginia, home to the Confederate capital, Richmond. The fact that its northern frontier, the Potomac river – with Maryland, and the federal capital, Washington, on the other side – was also the line between North and South, explains why so much of the war, at every stage, was fought there. In the west the field of battle was defined by the lower Mississippi river, and the southern tributaries of the Ohio river, which joined the Mississippi at Cairo, Illinois. Of these the most important was the Tennessee river, and on the river itself the most important town, strategically, was Chattanooga, in the south-eastern corner of the state of Tennessee; the state was also the scene of many decisive battles, of which that fought at Shiloh – and won by Union forces commanded by General Ulysses S. Grant – was the greatest battle of the whole war.[14]

Chattanooga, the 'mountain city', although west of the Appalachians, was the key railway junction in the supply line – absolutely essential to the Confederate armies – between Virginia and Georgia. The railroad line north-east from the city led to Virginia, following roughly the course of the Tennessee river, while that across the mountains to the south-east – which was much shorter – provided the link with Atlanta, Georgia, the hub of the Confederate railroad system. There were also lines leading north and west from Chattanooga, in the direction of the front line of battle – and for that reason essential to the Confederate supply chain.

By 1862, the second year of the war, the strategic importance of Chattanooga was clear to the Union forces in the west; when, therefore, James J. Andrews – a plausible character, with a questionable past – presented a scheme for taking out the link between Chattanooga and Atlanta, which was operated by the Western & Atlantic Railroad, the generals listened to him. The scheme was simple enough: at a convenient station on the railroad somewhere north of Atlanta, Andrews, with a force of some twenty volunteers – including two or three locomotive engineers – would steal a train bound for Chattanooga. Then, as they proceeded north, they would cut the telegraph lines and destroy the track behind them.[15]

The first part of the operation went more or less as planned. Andrews, who himself came from Kentucky – a slave state that had stayed loyal to the Union – recruited his men, one of whom was English, from three Ohio regiments. Meeting together for the first time, on Monday, 7 April 1862, at Shelbyville, a Tennessee town just inside the Union lines some 50 odd miles north-west of Chattanooga, the men, after a briefing from Andrews, split up into small groups to make their way on foot across the Confederate lines to Chattanooga. Appalling weather set them back, but even so all but two arrived in time to take a train – a day later than planned – to travel as ordinary passengers down the line to Atlanta. Their destination was Marietta, 20 miles short of the city, where they arrived to spend the night in a hotel on the evening of Friday, 11 April.

Except for two who overslept, the men, as planned, all caught the first train back to Chattanooga – the one they planned to steal – on the Saturday morning. They would travel as ordinary passengers to Big Shanty, 8 miles north, and the first scheduled stop down the line. There would then be far fewer people around to impede the operation, which proved to be the case. The engineer, the conductor and almost all the passengers left the train to have breakfast at the station, allowing Andrews – with two of his men who were trained engineers – to leave on the opposite side, and uncouple the last three coaches, two of which were for passengers and one for mail. This left a train consisting of a locomotive named 'The General', its tender full of wood fuel and three empty box cars. At a signal from Andrews, the two engineers mounted the footplate, while the remaining men stepped out of the passenger coaches, walked a few yards down the track and boarded the box cars. The line ahead was clear and the much-shortened train moved off without anyone around realizing what was happening. Such was the beginning of the 'Great Locomotive Chase'.

Although the distance between Big Shanty and Chattanooga was only 110 miles, a long and difficult day was ahead of the men who had stolen 'The General'. They had above all to take into account trains coming in the other direction on a single-track railroad. Andrews, having acquired a timetable, knew in advance the stations where trains could pass each other on parallel tracks; in particular, he knew that he would have to wait at Kingston, 31 miles out of Big Shanty, for a southbound train. In 1862 running speeds were still very slow, so that 'The General' – with a number of stops for minor adjustments to the engine, pulling up the track behind them and cutting telegraph wires – needed two and a half hours to cover this distance. At Etowah, 15 miles out of Big Shanty, they passed, for the first time, another locomotive, the 'Yonah', working a short branch line leading to local iron works. Unwisely the raiders left it alone.

The real troubles started at Kingston. For one thing Andrews had to explain to the station staff the arrival of an odd unscheduled train, with men on the footplate they had not seen before – the regular train crews were known all along the line. Andrews' story was that his train had been organized at short notice to bring urgent supplies to the front line beyond Chattanooga; this was just about credible, because the Union armies had in the previous days captured considerable Confederate territory in Alabama, not far to the west of the Western & Atlantic line to Chattanooga. Although Andrews had got away with his story, he was dismayed to see that the scheduled southbound train, when it arrived at Kingston, had a red flag on the last carriage – indicating that a second, unscheduled train was following on behind it. But when the second train arrived, it too had a red flag, so Andrews had to wait for a third train; when he was finally able to proceed north from Kingston, he had been there for more than an hour. This was critical, because sooner or later the railroad men left behind in Big Shanty would organize a pursuit.

The theft of 'The General' registered almost immediately to those left behind in Big Shanty. Impetuously, three railroad men, including the train's conductor, William Fuller, ran after it – a measure of desperation that in the end paid off. After 2 miles they encountered a maintenance and repair crew, with a hand car used for carrying material and equipment – and propelled along the track with poles. Fuller then appropriated this for continuing the chase at a somewhat higher speed, hoping to find the 'Yonah' at Etowah. Luck was with him, and with the 'Yonah' he covered the 16 miles to Kingston in as many minutes. Even at this breakneck speed he was still too late for 'The General'. With the track through Kingston station blocked by the three trains from the north, Fuller had to switch to another locomotive, the 'William R. Smith', that was due to take a train down a branch line to the west of Kingston. This could only take Fuller 3 miles to a point where Andrews and his men had broken up the track. Once more there was nothing for it but to continue on foot: this was by no means hopeless, since Fuller knew that a southbound freight train was due to come down the line.

In the meantime, Andrews, with 'The General', had passed this train at Adairsville, some 10 miles north of Kingston. Its engineer, with a locomotive called the 'Texas', talked to Andrews, and told him he should wait, along with the 'Texas', for a long overdue southbound passenger train. Andrews, knowing better, persuaded the engineer of the 'Texas' to continue his journey south, so unblocking the line north of Adairsville. Recklessly, Andrews chanced reaching the next station, Calhoun – some 9 miles down the line – before the passengers had left it. Travelling at more than a mile a minute, with its whistle blowing almost continuously, 'The General' was just in sight of the station as the passenger train was pulling out. The engineer of the latter prudently reversed it into the station, allowing 'The General' to pass alongside it into a siding. Once again Andrews had to use his story to persuade the other engineer to continue on south, so that 'The General' could leave to the north.

By this time Fuller, running north along the tracks, had met the 'Texas' travelling south. Once the engineer heard from

Fuller what was up, he willingly joined the chase. Running the 'Texas' in reverse, he first shunted all the freight cars into a siding at Adairsville, and then, with only the tender, put on speed so as to catch up with 'The General' as soon as possible. Passing the southbound passenger train at Calhoun, the engineer of the 'Texas' caught sight of 'The General' just as the raiders were preparing to set fire to the long wooden bridge over the Oostanaula river. With no time to do so, the only way open to them was to continue northwards without delay. The only alternative for Andrews, with twenty armed men, was to stand his ground at the entrance to the bridge, and if necessary engage in battle with the railroad men on the 'Texas'.

The decision not to do so was fatal, just as that taken earlier in the day not to disable the 'Yonah'. Firstly, once the 'Texas' reached Dalton, some 18 miles further down the line, it was finally possible to telegraph the news of the capture of 'The General' to Chattanooga. This meant that a train was immediately sent up the line to confront 'The General'. Secondly, 'The General', just beyond Ringgold, and only 20 miles short of Chattanooga, ran out of fuel. It had loaded a cord of wood between Etowah and Kingston, but this would never get the train to Chattanooga. The end was inevitable. Andrews and his men had no choice but to abandon the train, leaving each one of them, on his own initiative, to make it back to the Union lines. Not one succeeded: the men, once captured, were tried and convicted as spies. Andrews and seven others were hanged. The remaining men, with their fate uncertain, ended up in prison in Atlanta, from where eight eventually escaped, all making it back to the North along a variety of routes. The few then left in Atlanta were later freed as a result of an exchange of prisoners. On 23 March 1863, at a ceremony at the War Department in Washington, six of the raiders became the first recipients of the newly instituted

Medal of Honor. They then walked the short distance to the White House, to be received by the President Abraham Lincoln. Better still, they were all granted sixty days' leave to return to their homes in Ohio; it is appropriate to the theme of this book that they travelled by the B & O railroad, the earliest in the United States.

Far to the west, the two best-known generals at the end of the war, William T. Sherman and Ulysses S. Grant, both relied on railways for transporting troops to Vicksburg and Atlanta, where they won decisive battles for the Union. Following his victory at Atlanta – achieved after advancing down the railroad from Chattanooga which in 1862 had been the scene of the great locomotive chase – Sherman, by concentrating on the destruction of the remaining railways in Southern hands, had found the strategy that led to the Confederate General Robert E. Lee's final surrender at Appomattox in April, 1865.

8

BRITISH RAILWAY MANIA

The first shock of a great earthquake had, just at that period, rent the whole neighbourhood to its centre. Traces of its course were visible on every side. Houses were knocked down; streets broken through and stopped; deep pits and trenches dug in the ground; enormous heaps of earth and clay thrown up; buildings that were undermined and shaking, propped up by great beams of wood ... Hot springs and fiery eruptions, the usual attendance upon earthquakes, lent their contribution of confusion to the scene. Boiling water hissed and heaved within dilapidated walls; whence, also, the glare and roar of flames came issuing forth; and mounds of ashes blocked up rights of way, and wholly changed the law and custom of the neighbourhood.

In short, the yet unfinished and unopened Railroad was in progress; and, from the very core of all this dire disorder, trailed smoothly away upon its mighty course of civilisation and improvement.

Charles Dickens.[1]

The Stockton and Darlington Railway

Chapter 4 relates how, in the first quarter of the nineteenth century, steam locomotives proved their unrivalled utility for the transport of coal on the railways, which had long linked mines with local waterways. Although by this time there were altogether some hundreds of miles of mining railways – particularly, though by no means exclusively, in the north-east of England – there was no question, ever, of an integrated regional network. The most that was ever achieved was the sharing of lines by neighbouring mine-owners, who were also responsible for their construction and operation. Before the nineteenth century there was no concept of railways constituting a public utility offering transport facilities, even if only for a restricted market. The success of stagecoaches in providing for the transport of passengers on the newly constructed turnpikes – which by the end of the eighteenth century effectively constituted a nationwide system – had no parallel when it came to the transport of commodities by rail. On the contrary, the construction of the British canal system in the last quarter of the century would seem to have pre-empted any need for an alternative nationwide system for the surface transport of goods.

In the north-east, as old mines were worked out, while industry's demand for coal steadily increased, this could no longer be satisfied by coalfields whose location enabled mine-owners to provide for their own local surface transport. This, as shown in Chapter 2, favoured locations with ready access to two rivers, first the Tyne and then the Wear. A third river, the Tees – defining much of the boundary between County Durham and Yorkshire – was much less favoured, even though it was navigable by sea-going traffic to the port of Stockton, some 15 miles inland, which had pretensions to

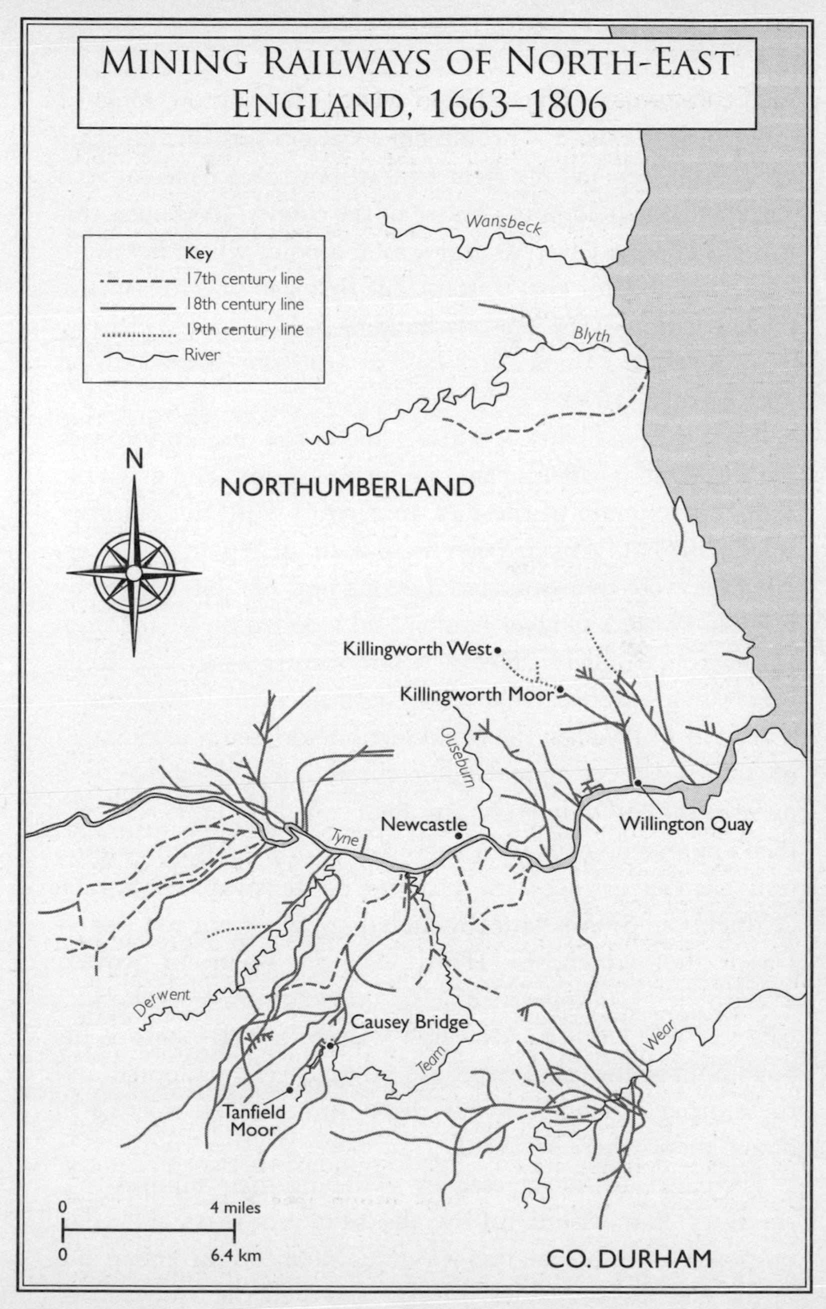

MINING RAILWAYS OF NORTH-EAST ENGLAND, 1663–1806
Key
17th century line
18th century line
19th century line
River
N
NORTHUMBERLAND
Wansbeck
Blyth
Killingworth West
Killingworth Moor
Ousebum
Newcastle
Tyne
Willington Quay
Derwent
Causey Bridge
Team
Wear
Tanfield Moor
0
4 miles
0
6.4 km
CO. DURHAM

Map 4

rival Newcastle-on-Tyne. There – and in Darlington, some 12 miles further inland – prominent local citizens, in the years after 1800, coveted the right to transport coal from the rich Auckland coalfield in the centre of the county. (Although this was close to the River Wear it was at a point where it was no longer suitable for coal traffic.) The first step was to improve the navigation of the Tees, by building a new cut: to celebrate its completion a dinner was held at Stockton Town Hall on 18 September 1810.

There the possibility of a direct line of communication with the Auckland coalfields came up for discussion, and given the economic climate of the day, in a straw vote the majority would probably have been in favour of a canal. Times, however, were changing, and a strong new faction, headed by Edward Pease, a Quaker banker and woollen merchant from Darlington, argued in favour of a horse tramway. For some years surveys were carried out by both sides; the issue came to a head in 1818 when the canal faction adopted a recommendation for a direct route from Stockton to Auckland that took no account of Darlington. An alliance led by Edward Pease then appointed, as its own surveyor, George Overton, whose unrivalled experience with the cast-iron tramways serving the coalfields in South Wales included constructing the Pen-y-Daren line which, in 1804, was the scene of Robert Trevithick's triumph with a steam locomotive related in Chapter 4. Overton, who had written shortly before his appointment that 'Railways are now generally adopted and the cutting of canals nearly discontinued',[2] left no one in doubt about where he stood.

The canal faction reacted by switching their support to a tramway that would follow the same route as they had proposed for the canal: this was tantamount to surrender, and in the end the two sides jointly presented the Stockton &

Darlington Bill, based on the route proposed by the railway faction, to Parliament – only to see it defeated as a result of the opposition of two substantial local landowners, Lord Eldon and Lord Darlington. Overton, who had never encountered such difficulties in South Wales, contained his exasperation and surveyed a new route acceptable to the noble landowners. This led to the Stockton & Darlington Railway Act receiving the Royal Assent on 19 April 1821.

The passing of the act left Edward Pease and his supporters in a commanding position, from which they offered George Stephenson – now joined by his son, Robert – a contract for providing steam engines for hauling the trains of coal wagons on the new railway. Because of the topography of the route between Auckland and Stockton designated in the Railway Act, the mode of operation planned for the new line was somewhat involved. At its western end, the Etherly and Brussleton inclines were judged to be too steep for locomotives, so provision was made for winding engines, with horses drawing the wagons along intermediate sections of track. Then, on the eastern side – from the foot of Brussleton Bank – two steam locomotives designed by Stephenson, the 'Locomotion' and the 'Hope', would take the wagons the rest of the way, over relatively flat country, to Stockton. These were duly constructed by the Stephensons' factory in Newcastle and brought laboriously by horse-drawn wagon to the new railway's New Shildon workshop in the middle of September, 1825.

The line was ready for traffic on 26 September, with the formal opening the next day. This started at Witton Park at the western end of the line, with ten loaded coal wagons from the nearby Phoenix pit ready to be drawn by horses to the foot of the Etherly incline. There the steam bank-engine successfully wound them up to the top, so that they could run down the other side under the force of gravity. Once they were down at the bottom, the process was repeated – with a wagon loaded with flour added to the train – up and over the Brussleton incline, but this time the 'Locomotion' was waiting for them at Shildon Lane at the foot of the bank on the east side. The locomotive was already coupled to a train consisting of the 'Experiment', a specially designed passenger carriage built for the occasion at the New Shildon works, together with twenty-one new coal wagons converted to carry passengers. Then, hitched to the wagons from the Phoenix Pit – which had already completed the first part of their journey – the train went on its way to Stockton. Tickets were allotted to 300 passengers, but almost as many again clambered onto the train, which after one or two mishaps – corrected by George Stephenson with the help of his brother, James – reached the junction with the Darlington branch line at midday. There six of the coal wagons from the Phoenix Pit were uncoupled so that their contents could be distributed to the poor of Darlington, while two new wagons – one carrying leading citizens from the town and the other, the Yarm town band – replaced them. Along the line tens of thousands had gathered to see the unprecedented spectacle.

At quarter to four this incredible train reached its final destination at Stockton Quay, to be greeted by a seven-gun salute – repeated three times – church bells pealing, the town band, supported by the Yarm musicians, playing patriotic music, and a cheering crowd estimated at 40,000. The guests of honour made their way through the general pandemonium to the Town Hall for an official banquet that lasted until midnight, when the last of twenty-three toasts was drunk to George Stephenson.

If, in September 1825, the railway age had opened in style, the new line was soon beset by problems to be solved at the

New Shildon shed. The work was done mainly by the local manager, Timothy Hackworth, since George Stephenson was already involved in another enterprise, while his son, Robert, had left to manage a new project in South America – where he stayed for three years. As to the challenge facing Hackworth, he had to replace the cast-iron wheels with a better design, improve upon the rudimentary braking and reversing systems, deal with valve and pressure problems, devise springs for the wagons – essential if the trains were not to chew up the track – and, at the same time, train men as drivers of a machine quite unlike anything they had seen before. These engine men were remunerated at a rate of ¼d (ie. a quarter-penny) per ton per mile of goods carried, out of which they had to pay the wages of their firemen and the costs of fuel and oil. This provided an incentive for speed at any price in a complex system in which 'horse-leaders' also had their part in moving the traffic, even on the section of line between Shildon and Stockton on which the locomotives operated. The fact that the line was single, with only four passing loops per mile, required an elaborate system of priorities: the rule was that passenger coaches always had the right of way, while with coal trains horses had to yield to locomotives – a principle made even more difficult to enforce by the practice, on downhill runs, of unhitching the horse and letting it ride, with the train, in a special 'dandy-wagon'. When it came to sheer truculence the engine men, once described by Robert Stephenson as 'not the most manageable class of beings', were nowhere in comparison with the horse-leaders. Conflict was inevitable and lost the railway profits it should have earned: the situation improved as the passing loops were progressively lengthened, to the point – reached in the early 1830s – that the railway was double-track along its whole length; indeed by this stage the horses and their leaders had had their day.

The Liverpool and Manchester Railway

Even before the Stockton & Darlington Railway opened George Stephenson was involved in a much larger project, the construction of a railway linking the seaport of Liverpool with Manchester – the centre of England's booming cotton industry – 30 miles away. The background was quite different: between the two cities the Bridgewater Canal, opened in 1761, and built by the third Duke of Bridgewater for the transport of coal to Manchester from his mines at Worsley, already provided a transport link. In the early nineteenth century this did not – at the rates demanded by the Bridgewater Estate – satisfy the requirements of Liverpool businessmen concerned to transport raw cotton, imported from the southern states of America, to the mills in Manchester. The Manchester mill-owners felt the same way. A railway was the obvious alternative to the canal, and who could be better to build it than George Stephenson? As director of the whole project, the new Liverpool Railway Company appointed Stephenson but with his first task, the survey of possible routes for the new railway, he immediately ran into considerable local opposition. This was led by Captain Bradshaw, the managing trustee of the Bridgewater Estates, supported by Lord Derby and Lord Sefton. Since no railway could be built that avoided the land they owned Stephenson had every reason to be discouraged. His problems were exacerbated by strong local support for his opponents, so that those working for him on the survey were continually harassed.

Even without such problems it would have been difficult to find the best route for the proposed railway. Although Liverpool and Manchester were only 30 miles apart, there were a remarkable number of natural obstacles in the country that separated them. The civil engineering problems involved in the route finally advised by Stephenson were severe, but they would have been even more so on any alternative route. A tunnel longer than a mile had to be dug at Edgehill and a deep cutting at Olive Mount, while a viaduct had to be built across the Sankey Canal; worst of all, at least in the opinion of the railway's opponents, was the need to cross the swampland of Chat Moss, just short of Manchester.

By the time Stephenson's plans first came before the Parliamentary Committee considering the bill for the construction of the railway, his opponents – led by Captain Bradshaw, and represented by one of England's leading barristers – made mincemeat of his evidence. Counsel's words were scathing: 'Was there ever any ignorance exhibited like this? Is Mr Stephenson to be the person upon whose faith this Committee is to pass this Bill involving property to the extent of £400/500,000?'[3] These were hard words for a man long regarded as obstinate, jealous and egotistical;[4] although Stephenson could not deny the errors in his survey, he was only too ready to attribute them to the mistaken policies of his business partners. He did, however, learn his lesson, and when the bill was presented a second time it went through to become the Liverpool & Manchester Railway Act of 1826.

Stephenson and his business partners were by no means out of the wood. After being granted, in 1827, an Exchequer Loan of £10,000, they found it necessary, at the end of 1828, to apply for a much larger sum: the Loan Commissioners, not unnaturally, required a survey of the works already completed, and for this purpose they employed the 72-year-old Thomas Telford – perhaps the greatest civil engineer in British history. Telford, who described his inspection 'upon a line of thirty miles in the present unfinished and incomplete state of the works' as a 'tedious and laborious task',[5] was little impressed by the progress already made. At this late stage it was not even clear whether the line would be worked with horses, winding engines or locomotives – or, as in the early days of the Stockton & Darlington Railway, some combination of these three possibilities. As expected Telford's report was not very favourable, but he was persuaded to advise positively after the Company had promised a firm decision in favour of one form of traction and Stephenson had signed an affidavit binding himself to complete the railway in 1830.

Twice, in 1828 and 1829 – by which time Robert Stephenson had rejoined his father after his years in South America – the Liverpool & Manchester Board sent deputations to observe the workings of steam locomotives in Darlingon, Shildon and Newcastle. To the dismay of the Stephensons, both deputations advised in favour of fixed haulage by winding engines, but Robert held fast writing: 'locomotives shall not cowardly be given up. *I will fight for them until the last*. They are worthy of a conflict.' Then, having broken the news in a letter to Hackworth at Shildon, he received a classic and heartening reply:

> I hear the Liverpool Company have concluded to use fixed engines. Some will look on with surprise ... Do not discompose yourself my dear Sir: if you express your manly, firm, decided opinion, you have done your part as their adviser. And if it happens to be read some day in the newspapers – 'Whereas the Liverpool and Manchester Railway has been strangled by ropes', we shall not accuse you of guilt in being accessory either before or after the fact.[6]

In the end, in 1830, this led to the publication of a pamphlet entitled *Observations on the Comparative Merits of Locomotive and Fixed Engines*, but by this time the

Stephensons had – with the crucial support of its treasurer, Henry Booth – won round the Railway Board, which, taking into account expected 'improvements in the construction and work of locomotives', decided, on 20 April 1829, to offer a prize of £500 for the best locomotive. If, then, this produced a winning entry of such merits as to make clear the superiority of locomotives, the issue would be decided in favour of this form of traction.

The competition produced hundreds of entries from across the world, but at the end of the day only three seriously counted. These came from the Stephensons, father and son, in Newcastle, Hackworth in Shildon and the firm of Braithwaite & Erickson in London, who had built up a reputation for steam carriages running on roads. On 31 August 1829 the Liverpool & Manchester directors resolved 'That the place of tryal for the Specimen Engines on the 1st October next be the level space between the two inclined planes at Rainhill; and that the Engineer prepare a double Railway for the two miles of level, and a single line down the plane to the Roby Embankment.'

Competing locomotives had to meet elaborate specifications. They must be mounted on springs and weigh not more than 6 tons with water if carried on six wheels, or not more than 4½ tons if carried on four wheels. They must consume their own smoke. A 6-ton locomotive must show itself capable of drawing 'day by day' a gross load of 20 tons at 10 miles per hour, a 5-ton locomotive 15 tons, and so on in proportion to weight. Steam pressure was not to exceed 50lb per square inch, but the Company reserved the right to test the boiler up to 150lb hydraulic.[7]

In 1828 the Stephensons had set up new works in Forth Street, Newcastle, where they developed the prototype of the modern locomotive. There, in the summer of 1829, all were busily engaged on constructing a new model to compete in the Rainhill trials. By early September it was ready for trials at Killingworth where George Stephenson had first made his name at the beginning of the century. These were so successful that Robert Stephenson, a born pessimist, was able to advise Henry Booth that 'On the whole the Engine is capable of doing as much if not more than set forth in the stipulations.'[8]

On 12 September the locomotive was dismantled and loaded on to horse-drawn wagons for the cross-country journey by road to Carlisle. There it was transferred to a canal lighter for transport by water to Bowness, where it was transhipped on to a steamer bound for Liverpool, to be transported, on the final stage of its journey, by wagons. These then brought the dismantled engine to the Crown Street railway workshops for reassembly. At some time in the course of these few days the locomotive received the name 'Rocket' with which it has gone down in history. It is also worth noting how, for carriage overland, it had to rely on two modes of transport – horse-drawn wagon and canal boat – that would be almost totally superseded by the revolution it was to herald.

The Rainhill trials were finally held five days late on Tuesday, 6 October 1830. The scene along the prescribed course would have been appropriate to a racecourse, with a grandstand at the mid-point and thousands of spectators along both sides of the track. There were five runners, listed as follows:

1. The 'Novelty', entered by Messrs. Braithwaite and Erickson of London.
2. The 'Sans Pareil' entered by Mr Ackworth [*sic*] of Darlington.
3. The 'Rocket' entered by Mr Robert Stephenson of Newcastle-upon-Tyne.

4. The 'Cycloped' entered by Mr Brandreth of Liverpool.
5. The 'Perseverance' entered by Mr Burstall of Edinburgh.

The last two entrants were not serious competitors. The 'Cycloped' was in any case disqualified since its engine was a sort of treadmill operated by horses; it was only allowed to enter because its owner, Mr Brandreth, was a member of the railway board. Following an accident during transport en route the 'Perseverance' was so late for the trial that its owner, Mr Burstall, noting how well its rivals had performed, decided to withdraw it. This left the 'Novelty', the 'Sans Pareil' and the 'Rocket' in the field.

Although the performance of the competing locomotives was to be measured over a distance of 1½ miles – marked out by white posts at each end – the actual track extended to terminal points one eighth of a mile beyond, where the engines could take on fuel and water, and if necessary be repaired. For every competing locomotive the trial was divided into two halves, in each of which it was required to haul a train of stone-filled wagons at three times its own weight ten times in each direction along the course. This ensured that the total distance travelled in each half would be equal to that between Liverpool and Manchester. Going east – towards Manchester – the locomotives were to pull the wagons; going west, towards Liverpool, they were to push them. The time-keepers were to measure not only the times taken to run the actual course, but also that between runs; fuel and water consumption were to be measured with equal care.

The 'Novelty', which was the crowd's favourite, started spectacularly on its first run, reaching a maximum speed of 38 m.p.h. and covering a mile in 1 minute and 53 seconds. Nemesis, however, soon overtook it. An explosion caused by a blow-back from the furnace put the 'Novelty' out of action.

The judges then called up the 'Sans Pareil', but Hackworth, confronted with a leaking boiler, was not ready. By this time it was raining heavily, and the judges adjourned the contest until the following day, when the 'Rocket' would show its paces.

Robert Stephenson's locomotive weighed in at 4 tons, 5 cwt – well under the prescribed limit – and in just under an hour, with its steam pressure built up from cold to 50lb, it was ready to be coupled to two wagons loaded with stone and start its first lap. Stephenson prudently decided not to use full power, although he allowed it to increase during the last three laps on the first half of the trial. Having completed this without any mishap, the 'Rocket' was ready for the second half; this went so well that in the last lap – towards Manchester – the driver gave it full regulator, to achieve, with a time of 3 minutes, 44 seconds, an average speed of just over 24 m.p.h. For the full course of 60 miles the 'Rocket's' average speed was just under 14 m.p.h. – safely beyond the 10 m.p.h. prescribed by the trial rules.

Although the judges reported to the directors that the 'Rocket' had demonstrated 'in a very eminent degree the practicality of attaining a high velocity even with a load of considerable weight attached to the engine', the owners of the two competing locomotives proved to be poor losers. To satisfy them the trials were extended into the following week, but this did them no good: neither locomotive came near to completing the prescribed 60-mile course. Even so the public was not entirely convinced, noting that in the trials the locomotives only had to perform on the level. George Stephenson took up the challenge, and in the days following the end of the trials had the 'Rocket' take wagons carrying up to forty passengers up and down the 1 in 96 Rainhill incline, just beyond the end of the trial track. After such demonstrations:

> there could be no more talk of horses or fixed haulage engines. The battle for the locomotive had been most decisively won and it was as an exclusively locomotive-worked railway that the Liverpool and Manchester went forward to completion.[9]

As a result of the trials the railway not only purchased the 'Rocket', but ordered six similar locomotives, all of which – with much improved designs born out of Robert Stephenson's experience with the 'Rocket' – were delivered in the summer of 1830. At the last minute a seventh locomotive, the 'Northumbrian', was added just in time for the opening ceremonies on 15 September 1830. Then, only three weeks later, a locomotive of an entirely new design by Robert Stephenson, the 'Planet', was added – to make all the first seven obsolete within two years.[10] A year later Stephenson delivered two new locomotives – 'Samson' and 'Goliath' – built on a variant of the 'Planet' design, to work the 1-in-96 gradients of the Whiston and Sutton inclined planes. On its test run the 'Samson' hauled 80 tons up the inclines and 200 tons on the level; little more was ever heard about winding engines, although in the Edgehill Tunnel that connected the Liverpool terminus to the Wapping docks on the River Mersey, a winding engine was needed to haul rail traffic up the steep incline.

The official opening of the Liverpool & Manchester Railway on 15 September 1830 was not the same happy event as that of the Stockton & Darlington some five years earlier. Part of the problem was that William Huskisson, the local MP, had persuaded the Prime Minister, the renowned Duke of Wellington, to be present as the guest of honour at a time when his studied resistance to any reform had made him most unpopular in England's northern industrial towns. His decision to attend was calculated to improve the local image of the Tory Party, but it meant, at the same time, unnecessarily elaborate preparations for the day's events.

These involved reserving one of the two parallel tracks for the Duke's train of specially built carriages, while the other was to be used for a supporting cast of other trains – one of which was drawn by the 'Rocket'. Just under an hour from Liverpool the Duke's train stopped at Parkside, enabling a number of its passengers – including Huskisson but not the Duke – to get out and stretch their legs. A minute or two later the 'Rocket' approached at speed down the other track; the men who had left the Duke's train scrambled to get back, but Huskisson – whose limbs were stiff as a result of having had to sit through the lengthy funeral of the recently deceased King George IV at Windsor – stumbled and fell, so that his thigh was crushed under the wheels of the 'Rocket'.

George Stephenson did his best to sort out the situation, which meant recasting the elaborate timetable in such a way that both engines and trains soon found themselves in the wrong places. The critical question was whether the Duke should proceed – against his own better feelings – to Manchester. The Borough Reeve, who was part of the Duke's party, persuaded him that if he did not do so, the mob would think that their Prime Minister had got cold feet at the last moment. (There was as yet no telegraph to transmit news of the accident in advance of the train.) The Duke agreed that there was 'something in that', and with considerable difficulty Stephenson sorted out the chaos, making it possible for the train to arrive in Manchester at 3.30 p.m. Impatient and hostile crowds had blocked the way over the final mile, and when the train finally came to a halt in the station it was mobbed. The Duke refused to leave his carriage, and the Chief of Police urged an immediate return to Liverpool. Then, once clear of the mob on the return journey, the route of the

Duke's train was blocked by four locomotives on their way to pick up trains in Manchester. There was no other way about it: the four had to reverse all the way back to Huyton, where the way was finally open to allow the Duke's train a free run. This meant leaving only three, instead of seven, locomotives in Manchester to bring the waiting trains back to Liverpool: the only solution was to couple them together, to make trains far longer than their engines were designed for. Although by the time they reached the Whiston incline two of the engines from Huyton were available to help, even with five engines between them the trains from Manchester could not make it up the Whiston incline. All the male passengers had to get out and walk, in darkness and in pouring rain, but once they were back in their wagons at the summit, the trains had an easy run over the Rainhill level and down the incline to the Liverpool terminus, where, at 10 p.m., cheering crowds were waiting to greet them. Among the railwaymen, everyone from George Stephenson downwards had kept his cool, and the day was saved, with poor Huskisson – who had died that evening – as its only casualty. The next day the first train ran according to the timetable, carrying 140 passengers, and the line is still in business. Incontrovertibly, the age of railways had arrived.

A Future Full of Problems

The forty-year period from 1830 – when the Liverpool & Manchester Railway opened – until 1870 was one in which railways were constructed on a scale that would never be repeated. By the end of this period well over half of the railway network had been constructed not only in Britain, but also in the leading industrial nations of continental Europe and in that part of the United States east of the Mississippi – which still contains its industrial heartland. By the end of this period the most critical problems relating both to the construction and operation of railways had been solved. Also – in the course of the 1860s – new industrial processes revolutionized the manufacture of steel, while the mains generation of electric power became a practical proposition. From the 1870s onwards, the safety and efficiency of railways were enormously enhanced by both these developments. At the same time, railways came into their own as the means for opening up and exploiting new territories – the American west, European colonies in Africa and Asia, Latin America, Australasia, China and Japan. This opened a new era in railway history, which is the subject matter of Chapter 10. This chapter, on the other hand, is concerned about how the foundations came to be firmly established in the heartland of early railroading.

If the success of the Liverpool & Manchester Railway was to be repeated across the country, the problems encountered in its construction would require considerable further attention, wherever railways were to be built – in any part of the world – in the coming years. First, there would always be legal and political problems: these went back at least as far as the early seventeenth century when – as related in Chapter 2 – Huntington Beaumont, in operating his new wagonway, came into conflict with the Strelley family. George Stephenson's confrontation with Captain Bradshaw in the late 1820s showed that the cause of railways was still fraught with difficulties. Once the principle of nationwide railway transport was established in the 1830s, the choice of routes still involved the necessity for private bills in Parliament, overriding the interests of landowners, while at the same time depending on local political support. The line chosen for many a route was then the result of the geographical distribution of local support (generally with an eye to commercial advantage) and local opposition (often born out of fear for the disruption likely to follow from railway operations). In this way, for example, Peterborough, whose leading citizens were not slow to see the advantage of trains for the transport of bricks – whose manufacture was the most important local industry – became a railway town, while Stamford – 14 miles to the west – oblivious to any such advantage, was left out.[11] Other towns, such as Cambridge, accepted the railway provided it was kept at a distance from the city centre. Still others had no choice and were left out in the cold when the route bypassed them by a margin measured in miles. Such was the fate of Marlborough, when I.K. Brunel decided to avoid the downs and route his Great Western Railway through Swindon (where he also established the main workshop). Although Britain led the way, hard-fought railway politics was by no means a British phenomenon – as witness the history of the American Baltimore & Ohio Railroad, recounted in Chapter 5. Indeed, long before the end of the nineteenth century it was clear that the construction of railways, anywhere, was a political event; this is a major theme of Chapter 10. The politics often had an international dimension: in 1834 Belgium became the first railway[12] state on the Continent because of its need to develop its natural resources in coal and iron without being dependent on transport by water – its two great rivers, the Schelde and the Maas, both reached the sea in the Netherlands, from which it had only become independent in 1830. In the Netherlands, on the other hand, transport by water was grist to the mill – as it still is. This explains why, by the 1840s, the length of railway line in Belgium was ten times that of the Netherlands. In 1845, however, the Dutch Railways achieved an unusual distinction: at the request of the Austrian physicist, Christian Doppler, a locomotive drew an open wagon, carrying several trumpeters, at speed past a station on the line from Amsterdam to Haarlem. Observers on the platform recorded how, as the train passed by, the pitch of the trumpets lowered immediately, as required by Doppler's theory: the 'Doppler effect' has proved to be extremely important for physics, even that of the stars.[13]

Quite apart from law and politics, there were considerable problems in civil and mechanical engineering. As to the former, the railway builders – as already shown above – could draw upon a wealth of experience in the building of canals, docks and turnpikes. Even so railways were much more demanding: the fact that lines had little tolerance for curves and gradients meant the construction of embankments and cuttings, bridges and tunnels on an unprecedented scale. What is more, the weight and speed of trains required the permanent way to be built to very high standards, at a stage in industrial history when material technology was still quite rudimentary. Before the revolution in steel-making of the 1860s, track consisted of short lengths of wrought-iron rail; not only did these constantly have to be replaced as a result of wear and tear, but the gaps between successive lengths of rail meant a very rickety ride, with inevitable stress to the springs of the rolling stock and discomfort for passengers. At a time when there was a fever of railway-building, all this required a labour force counted in tens, if not hundreds of thousands. The emergence of a new underclass of Irish labourers was a social phenomenon with an impact across the whole of Britain no less than that of recent immigrant communities at the turn of the present century.

As for mechanical engineering, the steam locomotive required an entirely new class of skilled craftsmen, both in the workshop and on the footplate. Here, for a long time, safety was extremely problematic, both for those who worked on the railways, and for passengers. Fatal accidents were all too common: in the two years of operation at the Stephensons' Forth Street works culminating in the construction of the locomotives for the Liverpool & Manchester Railway, George Stephenson lost both a brother and a brother-in-law in such accidents. On the tracks braking systems had long been inadequate, particularly if the brakes were to operate upon all wheels – a problem only solved with the introduction of vacuum-braking much later in the century.

Signalling, Safety and Traffic Control

Last, but by no means least, was the problem of safety and traffic control: the fiasco on the opening day of the Liverpool & Manchester Railway was proof enough of the need for a comprehensive solution. It was clear from the start that, since their routes are determined by the rails, not by the driver, trains cannot take evasive action or move to avoid one another, while, because of their considerable momentum, speed cannot be reduced abruptly nor stops made in the distance visible to the driver. Plainly some new means of control was necessary to deal with these factors.

The answer was to be found in the development of railway signalling. Essentially this meant a system, operated by railway staff at key locations, whereby a signal was given to an engine driver when it was safe to proceed further down the line. At an elementary stage this could be no more than a lamp or a flag shown to the driver – a system still used occasionally in shunting off the main line. If asked what the lamp- or flagman knew that was hidden from the driver, the answer was in what he had observed – and ideally recorded – of the progress of other competing traffic. He could also be certain that the points were set right for the train in front of him. Although in Britain, as early as 1843,[14] levers for operating signals and points were for the first time brought together in a central metal frame, installing such frames in signal boxes – from which the whole field of operations could be seen – was a slow process that only neared completion in the 1860s.[15]

As part of this process introduction of mechanical interlocking gradually became standard; with this system the operation of a lever was blocked, when by doing so a route already set up would be endangered.

With the introduction of the signal-box levers – colour-coded according to their function – points were operated by a system of fixed rods and signals by wires. Points, consisting of two heavy blades linked by a tie-rod, being much heavier to move, had to be located within a distance of 150 yards. The same physical effort was required of the signalman to move the points either way – indeed the determination of which position was open, and which, closed, was more or less arbitrary. A mechanical signal, for obvious safety reasons, was always designed so as to return, under the force of gravity, to the 'red' or 'danger' position: the wire that operated the signal, by bringing it to the 'green' or 'clear' position, counteracted the force of gravity, so that if it was broken, for whatever reason – including vandalism – the signal would automatically return to danger.

In Britain, and in much of the rest of the world, the application of this principle led to the introduction of the standard semaphore signal,[16] in which horizontal, indicating 'danger', is the default position. In the 'clear' position the signal is at an angle of 45° to the horizontal, in either the upper or lower quadrant according to design – the actual instruction to an engine-driver is the same in either case: on a long route both types of signal could well be encountered. At a very early stage, a distant signal – in Britain the familiar yellow and black fishtail – was introduced to give an engine driver advance warning for the position of the following stop signal. The standard distance between the two of a quarter of a mile was determined by the minimum braking distance of a train.

This system was sufficient for the local control of traffic within a range measured in hundreds of yards, but it was of no use when it came to giving permission for a train to proceed beyond it towards the next signal box down the line. Here train control relied on the strict adherence to timetables, on the principle that a train would not run into the one in front of it given a sufficiently long interval of time between them. As an additional safeguard, where a train, by force of unforeseen circumstance, came to a halt, the guard was required to walk back down the line, waving a red flag, or placing detonators on the track to give an audible signal of the danger ahead.

In the United States, where timetable operation – popularly associated with the image of a railway employee looking at a large pocket watch – was considered to work more or less satisfactorily, it was given up with reluctance. To a degree its shortcomings were overcome by appointing dispatchers, responsible for the central direction of the movement of trains, for each section of the line – so that Andrew Carnegie was the first dispatcher on the Pittsburgh Division of the Pennsylvania Railroad, in 1854. To work well, however, the system required the electric telegraph, which was only adapted to it a year or two later.

Although the full potential of the electric telegraph for railway operations was slow to develop, the actual invention dates back to the 1830s, by which time scientific discovery earlier in the century had laid the foundations for the technology. The year to start with is 1800, in which Alessandro

Volta, an Italian physicist, invented the prototype of the modern battery. Surprisingly – given their subsequent history – batteries, for more than thirty years, were used almost exclusively for the purposes of scientific research. This, in turn, depended upon an extremely useful process – developed by Sir Humphrey Davy and known as 'electrolysis' – which enabled highly reactive metals, such as aluminium, magnesium, calcium, sodium and potassium to be isolated from the compounds in which they invariably occurred in nature.[17] This was a revolution in chemistry with very great importance for the material sciences – including particularly metallurgy – as they developed in the course of the nineteenth century. It hardly need be said that every new discovery worked for the benefit, one way or the other, of railway engineering – which, throughout the nineteenth century was always at the cutting edge of applied science.

In 1813, Davy, who was the president of the Royal Institution – which had rapidly become the most important centre for scientific research in Britain – appointed as his assistant a young man of twenty-two, Michael Faraday, who was destined to become one of the great scientists of all time. In particular he established the connection between electricity and magnetism, showing – at an early stage in his experimental career – how transmitting an electric current through a length of wire, wound round a core like a thread round a bobbin, created a magnet with its poles at the two ends of the core. Faraday's apparatus, which he called, appropriately, a 'helix', but is now known as a 'solenoid', was the prototype electric magnet.[18] With this one experiment he found a new use for the electric battery, when the two ends of the wire were connected to the electrodes. What is more, reversing the direction of the current – a result simply achieved by a switch – also reversed the polarity of the magnet.

In the early 1830s Charles Wheatstone, a friend of Faraday's, saw the potential of the solenoid for sending a signal down a length of wire over an indefinite distance limited only by the cumulative electrical resistance of the wire – which, to a degree, could always be overcome by using a more powerful battery. In the late 1830s this provided the basis for the electric telegraph, developed by Wheatstone, assisted by a colleague, W.F. Cooke, who quickly realized its potential for railway operations.

The actual telegraph was first demonstrated to the public, in 1837, over a mile-long line adjacent to a section of the new London & Birmingham Railway Company, and a year later, Brunel – realizing its potential use by his Great Western Railway – arranged for a trial along the 13 miles of railway between Paddington and West Drayton, a line, which, in an improved form, was extended to Slough in 1843. At this stage the only signal that could be sent down the line was visual, in that its message was conveyed by a rapid series of compass needle movements activated by a solenoid: these according to a simple pre-arranged code could then be read off the receiving instrument by a trained operator. This manner of operation, combined with a key for reversing the current in the line so that the compass needle could point both left and right, was the basis for the electric telegraph. In the United States Samuel Morse had also developed an electro-magnetic telegraph during the 1830s, and for this, in 1837, he filed his patent for the 'Morse' code – which, by assigning every letter of the alphabet and every numeral its own combination of compass needle positions pointing left and right, allowed any text message to be sent down the line. (In its later development the code was used for the familiar audible signals based on dots and dashes, but this required the development of a 'buzzer', which did not exist in the early days of Morse's telegraphy.) Morse, however, was critically short of funds, and it was only after the US Congress granted $30,000 for the construction of a telegraph line between Washington and Baltimore – along the route of the Baltimore & Ohio Railroad – that the breakthrough came. In May 1844 Morse sent the first telegram in history, reading 'What hath God wrought'. His words were prophetic: within a few years electric telegraphy, making possible instantaneous long-distance communication, was served not only by countless domestic lines, but by an international cable network. It was closely tied to steam-powered transport in two ways: on one side, without such transport – particularly at sea (as shown in Chapter 12) – the lines could never have been laid; on the other, the electric telegraph was to provide the essential means for regulating traffic – particularly by rail.

Given the interest at the end of the 1830s of Brunel and others in the electric telegraph, and Morse's success during the 1840s, it is perhaps not surprising that W.F. Cooke – after playing a key part in constructing the lines along Brunel's Great Western Railway – saw its potential for adaptation to instruments designed, specifically, for the control of trains. In 1842, while working for the London and North-Western Railway, he published a pamphlet entitled *Telegraphic Railways*, in which he suggested the principle of working which sooner or later was applied throughout the British rail network.

By this time some order had already been brought into the chaos of train control, which during the 1830s was largely left to the newly constituted Railway Police,[19] who had many other tasks to perform.

Although from the 1840s onwards centralized control from signal boxes – with specially trained signalmen recognized as skilled craftsmen – greatly enhanced both the safety and efficiency of railway operations, such control could not extend beyond the area in which the points and signals operated by any given signal box were located. There was therefore a sort of railway no-man's-land consisting of all sections of line beyond the range of signal-box operations: since this, for operating points, was 150 yards, and for signals, where the limit was determined more by limitations on the line of sight, some 350 yards, the no-man's-land accounted for the greater part of any railway system. Here the engine driver, once he had passed the last signal, was on his own – the safety of his train being dependent upon sufficient intervals of time between successive trains along the same line. This was all very well if timetables were adhered to, but there were any number of reasons for a train not doing so, whether it was a mechanical failure of the engine or an obstacle, such as a fallen tree, blocking the line. In such a case it was left to the guard to warn any train following on behind by placing detonators on the line or walking down it waving a red flag. One difficulty about this rule was that there was no way a guard could communicate with the engine driver, who was bound never to leave the footplate. If a train made an unscheduled halt, the guard, therefore, had no way of knowing the reason for it. Given the uncertain performance of early locomotives, a halt could well be the result of a mechanical failure, which the engine driver could put right in a matter of minutes. If, during this time, the guard was walking back down the line with his red flag and detonators, the train could move off without him. A guard had every reason, therefore, to temporize when even a short delay could lead to a serious accident.[20]

The electric telegraph was used to regulate and safeguard traffic outside the domain of a signal box in two different ways, which, in the course of time, coalesced into a single comprehensive system of traffic control. One way was to use bell signals to offer and accept trains between signal boxes. This was first suggested, in 1851, by the Telegraph Superintendent of the South-Eastern Railway, who wanted a means for describing trains over some 850 yards of railway line been Spa Road – which opened in 1836 as the first ever (but now long closed) railway station in London – and London Bridge. The underlying reason was the fact that at London Bridge trains could continue their journey over three different lines, so that the pointsmen could well use advance information about their destinations. The system worked with single-stroke bells, the clapper being operated by an electric magnet on current transmitted over a telegraph line. A signal was then acknowledged by its code being repeated back to the sender: where the signal described a train, such acknowledgement meant acceptance, necessarily implying that the intervening section of railway line was clear. The appropriate lever was pulled in the signal-box and the engine-driver, seeing that the signal before him was no longer at danger, knew that his train was free to depart. After the short link at London Bridge the system was next installed along 11 miles of the North Kent line of the South-Eastern Railway in 1851; by 1863 some 330 bells covered 275 miles of its lines.

The second use of the electric telegraph was for operating purpose-designed block instruments – such as had been envisaged in Cooke's pamphlet of 1842. The concept of the block was fundamental: this was essentially the no-man's-land beyond the control of the lever-frame in a signal box. The guiding principle was that for any train to enter the block, it had to be accepted by the signalman in advance of the train operating a switch on his instrument. In the early stages during the 1840s the switch – which took the form of a handle on a horizontal axis turning in a vertical plane – had only two positions, 'train in' and 'train out', operating a compass needle on identical dials in the two signal boxes at either end of the block.

In 1852 Edward Tyer, by patenting an instrument to be used in conjunction with bell signals, laid the foundations of an integrated system, which was first installed in 1855 between London and Rugby – a distance of 83 miles – with an average block length of 2½ miles. The short blocks would permit a headway of about five minutes for goods trains, and even less for passenger trains, so heavier traffic could be handled than under the time interval system. In the early years railways tended towards a standard block length of some 2 miles: at this stage managers failed to realize that no standard was necessary, so that where traffic was intense the system could operate with much shorter blocks. In the course of the next twenty odd years the system became standard in Britain, with adaptations in parts of France; in the United States on the other hand it was adopted – and then only in 1876 – by the Pennsylvania Railroad. Similar systems were also introduced in Germany and the Netherlands in the early 1870s, by which time railway managers in these two countries had the benefit of twenty years of knowledge accumulated in Britain.

The standard British system, in its final form, was based on a three-position block instrument. With two signal boxes in adjacent blocks, the controlling instrument was in the one receiving trains along a given line: the control consisted of a handle which turned a pointer to any one of three positions, labelled, respectively, 'line blocked', 'line clear' and 'train on line'. The handle also operated a two-way switch in a telegraph wire linking the instrument to the signal box offering trains along the line. Both instruments had identical dials, in each of which a compass needle could point to three positions, labelled in the same way as those indicated by the pointer of the controlling instrument. 'Line blocked' was the central default position prevailing when no current was sent along the telegraph line, an essential safety feature. The other two positions were determined by turning the pointer, left or right, so as to send an electric current to operate the compass needle at the other end of the telegraph line. This simple system ensured that any one of the three possible indications, whether it was 'line blocked', 'line clear' or 'train on line', would be the same in both signal boxes.

The block system was essential for maintaining a key principle of railway operation: no train can proceed without a road being made ahead of it. Where, at any stage in its journey, the home signal is clear, an engine driver knows that the block ahead is clear at least as far as the next home signal. Moreover, a distant signal will give advance notice of whether he can proceed beyond it into the next block. The necessity for such a rigorous system is determined by two factors, both inherent in railway operation. The first is that the weight of a train means a long braking distance (to say nothing of the time needed to build up speed), and the second, that a train, confined to its rails, has no other means of taking evasive action to avoid an obstacle.

Essentially every part of a railway system belongs to one, and no more than one, block, with each block having its own signal box: in a busy part of the system the area of blocks will be small – so that trains will be continually moving across the boundaries between them – while, where traffic is light, blocks can extend for miles. In the operation of a railway the role of drivers is largely passive – so much so that driverless trains, in one form or another, ran, in particular cases, throughout the twentieth century. An example is the Post Office underground railway in London, which operated from 1927 to 2003, although it was preceded by similar railways in Berlin and Chicago. Since these railways depended upon electric traction the necessary technology was only being developed at the end of the nineteenth century. Since 1987 London's Docklands Light Railway has carried passengers in driverless trains over its own separate network. In the nineteenth century, however, drivers were essential to control the train, but in doing so they always followed the signals – taking into account the vagaries of the line, such as special speed restrictions, which they were expected to know in advance. The essence, therefore, of long-distance operations was for signalmen to pass trains on from one signal box to another down the line – taking for granted that drivers would obey the signals. Theirs, then, was the proactive role in train operation – as opposed to the reactive role of the drivers.

Unfortunately for early railway travellers it took some forty years odd before often reluctant railway managers recognized the inherent weakness of any system of traffic control depending on time intervals, or accepted that alternative systems based on distance between trains were entirely practical: the managers' state of denial was based not so much on their conservative instincts but on the compulsive need to avoid unnecessary expenditure at a time of cut-throat competition for traffic. Telegraph lines were a good thing when the traffic they carried was paid for; when, however, they connected block instruments the return was measured in terms of safety. After every serious accident – of which there were all too many – the debate about improved safety, and the costs attached to it, was very much in the public domain. The economic costs of accidents, arising both out of the damage to a railway's own track and rolling stock and out of the claims made for injury to the person and loss of goods, also provided a strong argument for investing in block instruments, which at the end of the day persuaded even the most conservative railway managers.

During this whole early period the Railway Inspection Department of the Board of Trade, which had been set up by Parliament in 1840, had conscientiously performed its statutory duty of investigating and reporting on railway accidents, and in doing so consistently pointed out what means could have been adopted to avoid them – as it still does today. By the end of this period the rule of 'lock, block and brake', which had become the inpectors' *leitmotiv*, required the installation of interlocking gear and block-working on all passenger lines – regardless of cost. In 1873 Parliament reacted by passing a new Regulation of Railways Act, but it was only in 1889 that its provisions became mandatory. Even so, the year 1870 – forty years after the opening of the Liverpool & Manchester Railway – can be taken as a watershed. By this time the public and Parliament were no longer willing to accept horrendous accidents such as had occurred in the past. At the same time new technology, in fields such as the generation of electricity and materials science, was opening the way to even faster, and considerably more comfortable trains.

The block system, as described above, is that of lines dedicated to traffic proceeding in only one direction: this was from very early days the norm for double-track operation, such as was established from the beginning on the busy Liverpool and Manchester line – a precedent followed on British mainlines almost everywhere. Single-track operation, common enough on British branch lines, and standard in parts of the world – such as the American West – where long distances had to be covered, required different purpose-designed instruments, but the overriding principle of making a road for every train still applied. Basically, the location of a signal box was determined to give the signalman a direct view of all junction operations involving switching points. Within this area the interlocking between levers in the signal box prevented conflicting roads being set up: outside it the position of trains was revealed by the state of the block instruments, which, if correctly used, would achieve the same result. The system was not 100 per cent foolproof. A signalman could fail to restore the signals after a train passed, while a driver could run through a signal at danger. The historical development of railway signalling is one of many different innovations – involving either new electro-mechanical gizmos or improved procedures – designed to reduce the possibility of such human errors. Others besides railwaymen were also at fault, such as wayfarers who failed to look out at an unguarded level-crossing, or drovers who allowed cattle to stray onto the permanent way. Poor maintenance of track, locomotives and rolling stock also took its toll. The lesson was learned at a very early stage: whatever precautions might be taken, trains were monsters with an unprecedented capacity to destroy life, limb and property – as noted by Charles Dickens in the passage quoted at the head of this chapter. Only the law could take things in hand, and this it did on a grand scale – almost everywhere that railways were built.

Law and Railways

The number of hits, which are the result of entering the keywords 'law', and 'railway' into a computer search of the catalogue of any copyright library are to be numbered in thousands. Significantly far more of them will relate to the period before 1850 than will to that after 1950. The earliest of the relevant books recorded in the Cambridge University Library was written by one C.F.F. Wordsworth and published about 1837. Its title, *The Law Relating to Railway, Bank, Insurance, Mining and other Joint-Stock Companies*,[21] refers to just one of the many branches of the law that was revolutionized following the introduction of rail transport.

The joint-stock company is a good place to start looking at the law as it relates to railways. Its history goes back to the incorporation of the East India Company in the year 1600, which was followed, until the early eighteenth century, by those of other great international trading companies financed and managed from London – such as notably the Hudson's Bay Company (founded 1671) and the Royal Africa Company (founded 1672). This line of development ended with the crash of the South Sea Company in 1720, which led to the Bubble Act of the same year, which made it extremely difficult to incorporate new joint-stock companies. To a degree this was closing the stable door after the horse had bolted, but for a hundred odd years British entrepreneurs adapted to the large-scale operations characteristic of the Industrial Revolution by resorting to unincorporated joint-stock associations – which had the considerable drawback that their legal existence was tied to individuals named as trustees or partners, on terms laid down in involved documents drawn up by conveyancers at the Chancery Bar. This meant that there were effective limits both to the number of members who could subscribe capital and to the period of time during which operations could continue.[22] Worse still, in the long run, there was no way of limiting the legal liability of members for the debts of such an association, for this would require it to become a legal entity distinct from its members – a possibility for which the general law did not provide.

Forms of business association acceptable during the eighteenth century were quite unsuited for railways as they developed after 1830: both the scale of investment and the interests of the general public required a form of incorporation in which shares could be widely held without any liability for the debts of the corporation attaching to the shareholders. Legislation was essential and in the United Kingdom, from 1834 onwards, a series of Acts of Parliament, culminating in the Companies Act of 1862, achieved this result. Although incorporation, as provided for in the legislation, was open to almost any form of business enterprise, it was the railways that led demand for fixed capital on the scale which it made possible. As noted by Simon Ville:

> by 1850 all of the largest firms listed on the Stock Exchange ... were railway companies ... the top fifteen companies accounted for 62 per cent of total paid-up capital in the UK, thus dwarfing manufacturing industry. The London and North Western Railway (LNWR) had raised more than £29 million by 1851, employed a work-force of 12,000, and operated 800 miles of track. This was a giant scale of operations for the time and these figures would still have outstripped most British manufacturers half a century later.[23]

After the 1830s, railways consistently accounted for more than a third of all new capital formation,[24] so that at the end of the nineteenth century half the capital value of the securities traded on the London Stock Exchange was attributable to them. (The figure is now (2007) less than 4 per cent.)

By the standards of any earlier age the most formidable legal problem facing railways in Britain was the acquisition of the land on which their lines were laid. Throughout the nineteenth century this depended on the long-established procedure of a private Act of Parliament, laying down the terms on which land could be compulsorily acquired. In the course of the century hundreds of such acts were passed, relating to different sections of line. Just as in the case of the Stockton & Darlington Railway described on page 148 they all required a hearing before a specially convened committee, before which both the railway interests and those opposed to them were represented by specially appointed parliamentary agents. Inevitably the secular trend during the nineteenth century was in favour of the bills being passed, but much depended on the strength of local interests represented before the committee. These included not only landowners (some of whom made handsome profits by selling to railway companies) but also those involved in other forms of transport – such as canal boats. There were also local residents who rightly feared loss of amenity following the construction of a new railway. However, in general, the broad mass of the population, as represented in every class of society, saw only too clearly the benefits, in the form of affordable and rapid transport, brought by railways. By the end of the 1840s their opponents knew full well that they were fighting a losing battle. It was not always so: successive Dukes of Argyll consistently – and successfully – blocked bills authorizing the construction of a railway along the Mull of Kintyre.

Another area in which Parliament intervened was in regulating both the rates and the safety measures adopted by railways. The landmark legislation was the Regulation of Railways Act of 1844, introduced by W.E. Gladstone as President of the Board of Trade. This contained a remarkable provision granting the state, after a period of twenty-one years, an option to purchase any new company earning 10 per cent or more. It imposed upon the railway companies an obligation to operate daily passenger trains at fares of 1d a mile: these, the so-called 'parliamentary trains', were intended, above all, to benefit workmen.[25] Gladstone's Act also provided for minimum rates for the carriage of goods (which were also laid down in the private Acts of Parliament). The legislation was consolidated in 1854 by the Railway and Canal Traffic Act, under which every railway was bound to provide, without favouring any particular person, 'all reasonable facilities for the receiving and forwarding and delivery of traffic'. All this was to be enforced by a new Railway and Canal Commission – a very early instance of a regulatory agency. As to safety there was little that the 1844 Act could do – given the state of the art technology at so early a stage in railway history – beyond providing the new Railway Department of the Board of Trade with a right of general supervision. Parliament passed several acts relating to the safety of railways during the nineteenth century – often as a reaction to a particularly horrendous accident – but it was not until the 1870s that standard safety devices became compulsory.[26]

In one area, however, Parliament did take action, not so much to prevent accidents, as to compensate their victims. By a quirk of the common law the surviving relatives of a person who died as a result of the fault or negligence of another had no civil claim to damages: as was said often enough, 'It is cheaper to kill than to maim.' This injustice was remedied by a series of Fatal Accidents Acts, of which the first was enacted in 1846 as a result of the unprecedented number of such accidents occurring on railways, whether to passengers, passers-by or employees. Under this Act certain specified classes of dependent relatives were granted a cause of action if the deceased, had he been merely injured, would himself have had one. Damages depended on the direct pecuniary loss suffered by the claimants. The act, however, was of little use to employees, for in 1837 Lord Abinger, Chief Baron of the Court of Exchequer, in a case before him,[27] laid down the rule that an injured man had no cause of action where the injuries were caused by the fault of a fellow employee, serving the same master: this monstrously unjust decision reflected the common attitude of the time to working men – as reflected by the way the Duke of Wellington was welcomed when he came to open the Liverpool & Manchester Railway. Things were no better on the other side of the Atlantic: the rule was adopted only four years later in a Massachusetts case in which the successful defendant was a railroad company – a decision widely followed in other states.[28] In the United Kingdom the so-called 'rule of common employment' was finally abolished in 1948,[29] whereas in the US this depended upon enactments passed over time by the separate states in which the rule applied.

Working men, whether employed in the construction or operation of railways, were also at a considerable disadvantage when it came to negotiating conditions of employment. In 1834, when the railway age was well underway, six agricultural labourers from Tolpuddle in Dorset were sentenced to seven years' transportation to Australia after they had established a local branch of the Friendly Society of Agricultural Labourers. Although the sentences were remitted in 1836, the case sent a clear message to working men, including those employed on railways. It was only in the 1890s that Parliament, concerned by the number of accidents attributed to the long hours worked by railwaymen, finally took steps to improve working conditions. The Railway Regulation Act of 1893, which provided for a maximum 60-hour working week and a minimum wage, was followed by two more Acts, in 1897 and 1900, which gave 'the government almost complete control of railway operation and hours of labour'.[30] In 1897, also, The Workmen's Compensation Act made employers – including railway companies – liable to compensate employees for injuries caused 'by accident arising out of and in the course of employment', without there being any need to prove fault or negligence. Significantly, in Germany, such legislation had already been introduced by Bismarck in 1884, and Britain did not wish to be left behind. Although by this time, also, railwaymen had their own trades union, the companies were always contemptuous of their activity.[31] This became a public issue with the judgement of the House of Lords in *Taff Vale Railway Co.* v. *Amalgamated Society of Railway Servants*,[32] which held the union liable for the damages to the railway suffered as a result of a strike organized by its officials. The effect of the notorious Taff Vale judgement was reversed by the Trade Disputes Act of 1906 – a landmark in labour legislation – but even so the railways could still refuse to recognize the unions. None the less membership of the Amalgamated Society of Railway Servants reached nearly 100,000 by 1907. It was only after a national strike in 1911 that the principle of collective bargaining was finally established.

9

MATURE RAILWAY SYSTEMS

The Technological Dimension

This chapter takes the year 1870 as a turning point in the history of rail transport. With the solution of the basic operating and engineering problems the technology existed to construct railways in almost any habitable part of the world: the process was largely complete in Western Europe – including Britain where it had first started – and in the US east of the Mississippi. If, by the end of the 1860s, expansion beyond these parts of the world was blocked, it was by political and economic factors rather than by any shortcomings in the state-of-the art technology. Even so, there were still important gains to be made in this field, which must be considered before looking at the new railway geography as it developed in the last thirty odd years of the nineteenth century.

While, in 1870, improved signalling, based on the signal boxes with interlocking frames and the electric telegraph, had made railways much safer and more efficient, improvements were still needed in both track and rolling stock. Here the revolution in the production of high-quality steels for industrial use – the result of the invention of Bessemer's converter and Siemens' open-hearth furnace – transformed railway operation. In 1857 the Ebbw Vale Ironworks rolled a 47-lb steel ingot into a short experimental length of double-headed rail,[1] which the Midland Railway Company then laid at a busy cross-over just outside its station at Derby, where the standard wrought-iron rails – because of heavy wear and tear – had to be replaced every six months. Ten years later, the new steel rail, over which some 500 trains a day passed, was still in good condition. Inevitably such rails became standard, although the process was slow – partly through lack of funds and partly because of insufficient manufacturing capacity.

Steel also transformed the civil engineering of railways, particularly when it came to the construction of bridges. The lesson was learnt after the Tay Bridge, opened as the pride of British engineering in 1878, blew down hardly a year later in a December gale, carrying a passenger train with it to the river below. This appalling disaster occurred to a bridge made of iron. When it came to building the Forth Bridge, with two remarkable spans 1,750 feet long, the engineers turned to steel produced by Scottish steel-makers using the Siemens open-hearth process.[2] The bridge, when it opened for traffic in 1890, was the largest in the world. Proverbially painting and repainting it is a process that never comes to an end: however that may be the bridge still carries the main-line railway north of Edinburgh across the Firth of Forth.

The use of steel also transformed the design of rolling stock. New steel alloys made possible not only much stronger boilers and cylinders for locomotives, but also rims, or 'tyres', for wheels: this was a considerable improvement because it eliminated the uneven wear that had restricted British workshops – though not those in the US – to making locomotives with only a single pair of driving wheels. With steel tyres both two and three pairs of driving wheels became standard for new locomotives, such as the familiar Atlantic[3] (4–4–2) and Pacific (4–6–2) classes still constructed in large numbers in the twentieth century. This transformation was essential to the production of the more powerful locomotives needed for the growth of traffic in the last thirty odd years of the nineteenth century, but it was only in the 1890s that carriages, built on a steel framework, first had such amenities as corridors and even electric light – although, in the US, the Pennsylvania Railroad only introduced electric lighting in 1902. As to the former the Great Western Railway, in 1892 – the year in which the conversion to the standard 4 foot 8½ inch gauge was complete – introduced the first British train with a corridor running its entire length. This meant that lavatories were for the first time accessible to third-class passengers, something regarded as 'a startling innovation on the Great Western, which never forgot that they were "lower orders"'.[4]

At this stage, also, two 2-wheel swivelling bogies, long used in America, were introduced for both locomotives and carriages. It was well into the twentieth century, however, before the small standard British goods wagons – with only four wheels on fixed axles – yielded to the American standard with two 4-wheel bogies.

With more powerful locomotives, with four or six driving wheels pulling much heavier rolling stock at the higher speeds made possible by steel rails, it became essential, in the 1870s, to improve both couplings and brakes – which had long been so defective as to be the cause of all too many often fatal accidents. The standard link-and-pin coupling, although easy to manufacture, was prone to fracture – with almost inevitably the hazard of a divided train – and also dangerous to operate. Shunters making up long goods trains were frequent casualties. Any number of new couplings introduced in the 1870s, such as the British buck-eye or the American Janney Automatic Coupler,[5] proved to be safer and much more reliable.

The application of a close-fitting metal shoe to one or more wheel rims has always been the only way to brake a steam train. The problem is how to operate the system. If the braked wheels are part of the locomotive, this is an operation controlled from the footplate by a system of linked metal rods. The same is true if the braked wheels belong to the last carriage or wagon of the train, in which case the guard is responsible for applying the brakes.

For more than forty years there was no better system, which meant that guards – brakemen in the US – had to become skilful in judging gradients and reading signals, so as to apply the brakes according to what they saw from their look-out projecting above the roof of the caboose. (This is functionally the same as the British guard's van, but in Britain the lookout window projected out of the side rather than the roof.) Long freight trains in the US often had additional 'brakemen' at intermediate points, but in spite of hand signals transmitted along the length of the train, from the locomotive to the caboose, coordination was inevitably far from perfect. With nothing to go by except such hand signals, train staff were expected to know in advance the exact real-time state of the road, including not only the location of each signal and every change of gradient but also any special information, applying only at a particular time and place and relating to such matters as temporary speed restrictions – generally because of engineering works – on the permanent way. In the course of time, as increasing attention was paid to training and testing drivers and guards – engineers and brakemen in the US – such matters became second nature to them. If this, according to the popular adage, was 'a hell of a way to run a railroad', there was no better alternative – at least until George Westinghouse came on stage at the end of the 1860s.

Westinghouse's talent for making inventions combined with his capacity for exploiting them is unequalled in industrial history. Born on 6 October 1846 he acquired his basic engineering skills in the workshop in Schenectady, NY, where his father made agricultural machinery. After joining the US Cavalry as a private soldier in 1862, the second year of the Civil War, he became, two years later, at the age of eighteen, Acting Third Assistant Engineer in the US Navy. With the end of the war in 1865 he went to college, to drop out after only three months when he obtained his first patent – for a rotary steam engine. He went on to invent an instrument which replaced derailed freight cars on the train tracks and then started a business to manufacture it. With his focus on railroads he certainly had a good eye, which proved itself once more when, in April 1869, he obtained a patent for his air brake. Three months later, aged twenty-two, he formed the Westinghouse Air Brake Company in July of 1869 with himself acting as president.

Air brakes operate by using compressed air to retract the brake shoes from the wheel rims. The system of compressed air brakes is integrated over the whole length of the train. At the beginning of a day's run, the compressor, located close to the footplate and operating with its own steam engine, must build up sufficient pressure to release the brakes – so allowing the train to start its journey. The engine driver applies the brakes by means of a manual control that releases the pressure. Except in an emergency the occasional light application of the brakes is all that is needed to control the train, which has a number of advantages – smooth running, low wear and tear on the brake shoes, and minimum loss of air pressure. Just as with building up speed, losing speed with a heavy steam train is essentially a very gradual process – taking place over several miles of track.

Westinghouse's standard air brake, which was patented on 5 March 1872, revolutionized rail transport: after the Burlington Railroad, which operated westwards from Chicago, had adopted it, others soon followed. A reliable way of stopping trains, or, in any case, reducing their speed, it made it safe for them to travel at much higher speeds on the new all-steel rails. Westinghouse, while continually improving on his invention, opened new companies to exploit it in Canada and Europe, so that it was soon used worldwide. Before the end of the nineteenth century legislation, such as the American Safety Appliance Act of 1893, made this – together with automatic couplings – mandatory for all passenger trains. By 1905, over 2,000,000 freight, passenger, mail, baggage and express railroad cars and 89,000 locomotives in the United States were equipped with the Westinghouse Automatic Brake.

Where the Americans led the way, the rest of the world followed, though the adoption of the standard American braked freight trains made up entirely of wagons with two 4-wheeled bogies moved very slowly. British railways continued to work with unbraked goods trains, with their small four-wheel wagons, until well into the twentieth century. Slow speeds were seen as a price worth paying for considerable savings in capital costs, particularly when the standard unbraked wagon was economical for small loads, such as might be collected at a siding on a country branch line. In relatively densely populated Britain this was an important factor at a time when there was no alternative transport. The small British wagon was also much more convenient for the mechanical loading and unloading of coal, which, for well over a century, accounted for more goods traffic than any other commodity. The rest of Europe tended to follow Britain rather than America, if only because much the same economic factors applied. Even so, long before the end of the nineteenth century, every railway could choose equipment from any number of suppliers – so much so that diversity rather than uniformity was the rule. The main limitation, as time went on, was the necessity to comply with a multiplicity of laws enacted to enhance railway safety – which, together with the latest state-of–the-art equipment, they did with considerable success.

Particularly in the US, freight transport was transformed in the second half of the nineteenth century by the use of refrigerated railroad cars. Although large-scale commercial refrigeration started with lager beer, first introduced in the 1840s by German immigrants with such familiar names as Busch,[6] Pabst and Schlitz, its greatest commercial success was in the meat-packing industry. With the end of the Civil War new railroads soon exploited the great expansion of ranching west of the Mississippi, with cattle driven off the range to railheads in such places as Kansas City for onward transport by train to Chicago. In 1864 a consortium of nine railroad companies purchased a 320-acre area of swampy land in south-west Chicago for $100,000 which became the Chicago Union Stock Yards, an enterprise designed to be the commercial link between east and west. Here, in 1867, George Armour opened his first plant, to become the leader among the great meat packers of Chicago. The industry was transformed in 1872 when, by using newly invented ice-cooled units to preserve meat, it became able to process cattle at any time of the year, rather than just in winter. In 1882, Gustavus Swift developed the first refrigerated railroad car designed for shipping processed meat, so that live animals no longer needed to be shipped to eastern markets. (First-generation refrigerated railroad cars were already operating in the eastern states before the Civil War, but almost only for the transport of dairy produce and seafood.) In 1867 J.B. Sutherland of Detroit became the first inventor to patent a refrigerated railroad car. Then, with the new railroads to the west coast – described later in this chapter – the refrigerated car proved equally well suited to the transport of grapes, peaches, pears, plums, apples and citrus from California and apples, pears, cherries and raspberries from the Pacific Northwest, to say nothing of peaches from Georgia and citrus fruit from Florida on the other side of the country. At this stage, however, there was no refrigeration plant in the actual railroad car: instead specially designed ventilation systems operated in conjunction with ice packed in containers inside the car.

When it comes to the technology available to railways much the most important development after 1870 was electricity supplied by self-exciting generators based on elementary models devised by Michael Faraday and operating according to electromagnetic laws that he, more than any other scientist of his generation, had established.[7] The commercial generators were developed in Britain by Wheatstone, who had worked with Faraday, and the immigrant German scientist, Wernher von Siemens. In the United States the key figures were Westinghouse and Thomas Edison – inventor, together with the Englishman, Joseph Swan, of the filament-based electric light bulb – who engaged in a hard-fought battle as to whether the new generators should produce alternating or direct current, a question whose answer depended in part upon which was more suitable for the new electric chairs to be introduced by the New York state penal system.

In the end the battle was won by Westinghouse and, in Britain, by S.Z. de Ferranti,[8] both of whom realized the enormous advantages of alternating current for the transmission of electric power over long distances – a point eventually conceded by Edison. Even so the political battle, in its early stages, was won by local interests, which – quite unconcerned about long-distance transmission – were content with the small-scale generation of direct current, which, in the long run, was a costly mistake.

Railways exploited these developments in the realm of electricity in many different ways. First, a central power supply made possible the continuous recharging of the batteries used in railway signalling; what is more, by incorporating in railway carriages a small generator connected to the axles by a belt drive, a battery maintained on continuous charge could supply the current for lighting them with Edison and Swan's electric filament lamps.

Where the first part of this chapter relates the advances in railway-related technology from the 1860s onwards, noting, at the same time, their essential contribution to safety, efficiency, increased loads and higher speeds, the second part relates the way that the world's railway systems expanded in the same period. There are any number of cases to consider, almost all relating to just one country, but they fall into two separate categories. The first, defined by those parts of the world which were into railways from the very earliest days, comprises Britain and Europe on the one hand, and North America – mainly the US – on the other. The second is defined by the construction of railways in the rest of the world: although, in some countries such as India, this had started before the 1860s, most of these new railways were built towards the end of the nineteenth century or even in the opening years of the twentieth. The historical process can most conveniently be taken to end with the year 1914, in which the First World War started.

New European Railways

Whereas in Britain only 2,000 miles of new line were constructed during the 1870s – indicating that the railway system was almost complete – in continental Europe some 37,000 miles were constructed, mostly in economically backward regions such as the Iberian peninsula, Italy and the Balkans. The most spectacular achievement was, however, the construction of the great rail tunnels under the Alps linking Italy to northern Europe. Although construction on the first of these, the Mont Cenis tunnel linking Modane in France with Bardonecchia in Italy, started as early as 1857, it only opened for traffic on 17 September 1871. This, remarkably, was eleven years ahead of time, an achievement made possible by much-improved technology – pneumatic drilling, electric detonation of explosives, and dynamite – developed during the 1860s.

With the completion of the 8½-mile-long Mont Cenis tunnel, together with that of the railway along the Côte d'Azur in the same year, France and Italy acquired two new rail links. Two railways also linked Austria to Italy, that over the Semmering Pass, and a second, which provided a direct route to Germany, over the Brenner Pass. The great challenge, however, was to open more direct rail links to the Rhine and Rhône valleys, and this meant building tunnels in Switzerland under some of Europe's highest mountains.

The first of the two Swiss tunnels was the St Gotthard, which made possible the double-track railway between Turin in Italy and Luzern in Switzerland. This was a joint Swiss, German and Italian project, with Italy – whose economy

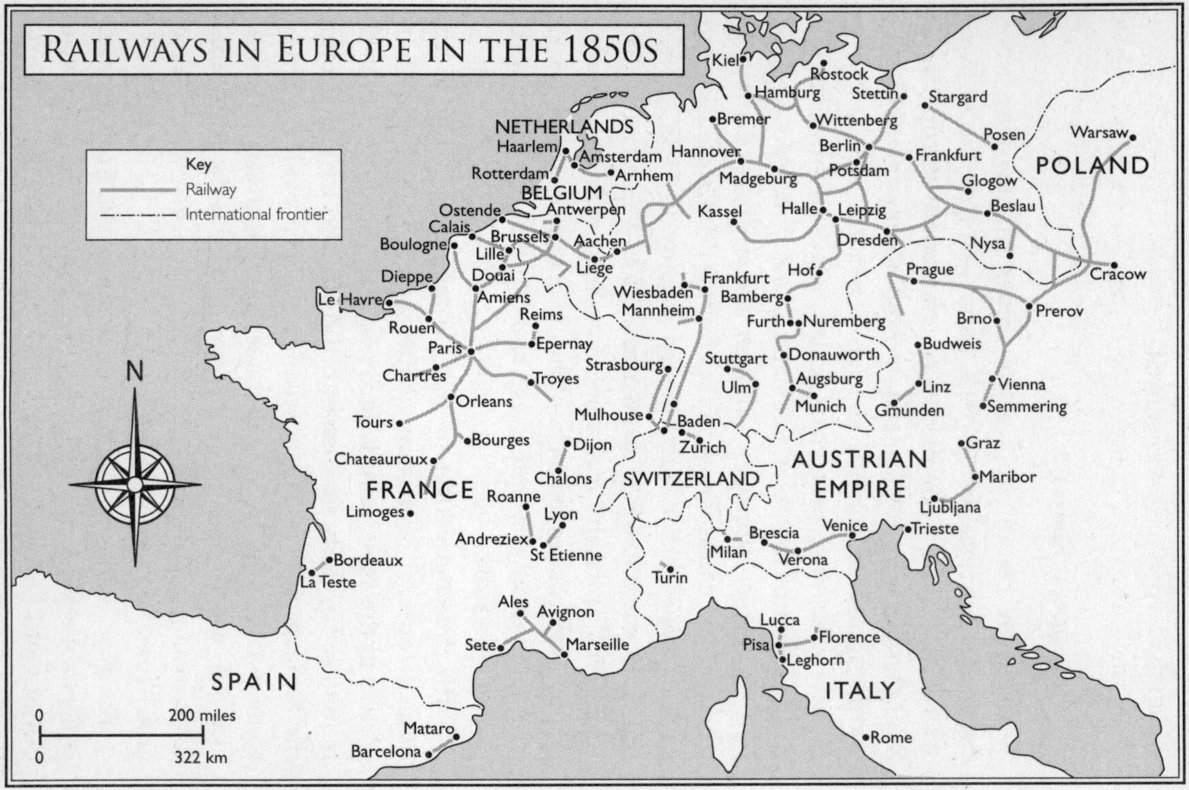

RAILWAYS IN EUROPE IN THE 1850S
Key
Railway
International frontier
N
0
200 miles
0
322 km
NETHERLANDS
BELGIUM
FRANCE
SWITZERLAND
AUSTRIAN EMPIRE
POLAND
ITALY
SPAIN
Kiel
Rostock
Hamburg
Stettin
Stargard
Bremer
Wittenberg
Posen
Warsaw
Haarlem
Amsterdam
Hannover
Berlin
Frankfurt
Rotterdam
Arnhem
Madgeburg
Potsdam
Glogow
Ostende
Antwerpen
Kassel
Halle
Leipzig
Beslau
Calais
Brussels
Aachen
Dresden
Nysa
Boulogne
Lille
Liege
Douai
Frankfurt
Hof
Prague
Cracow
Dieppe
Wiesbaden
Bamberg
Le Havre
Amiens
Mannheim
Prerov
Reims
Furth
Nuremberg
Brno
Rouen
Epernay
Budweis
Paris
Strasbourg
Stuttgart
Donauworth
Chartres
Troyes
Ulm
Augsburg
Linz
Vienna
Orleans
Munich
Gmunden
Semmering
Tours
Mulhouse
Baden
Bourges
Dijon
Zurich
Graz
Chateauroux
Chalons
Maribor
Roanne
Ljubljana
Limoges
Lyon
Andreziex
St Etienne
Brescia
Venice
Trieste
Milan
Verona
Bordeaux
La Teste
Turin
Ales
Avignon
Lucca
Florence
Sete
Marseille
Pisa
Leghorn
Mataro
Barcelona
Rome

Map 5

would benefit most from it – contributing more than half the estimated capital cost of 85 million francs. The agreement to build the tunnel was signed in 1871, with work beginning on 13 September 1872. The prime contractor, Louis Favre, who came from Genoa – a harbour city which stood to benefit considerably from the rail link through the tunnel – was unwise enough to commit himself to both a fixed price and a fixed time. The difficulties were horrendous, not only in constructing the tunnel, but also in finding a route for the railway to reach it at either end. On the southern side the Ticino Valley became so steep, narrow and precipitous as to make it impassable for any railway: between two small towns Giornico and Lavorgo a difference in height of 220 yards divided by a horizontal distance of some 3 miles, produced an impossible 4.4 per cent gradient. The only way round was to link the two towns with a spiral tunnel with two circles. This was not all: the same expedient had to be adopted – with a single circle – even higher up the line. Once at the site chosen for the tunnel mouth, Favre chose a Belgian method of tunnelling that proved to be quite unsuitable for the Alps. In 1876 Favre ran out of funds, so further progress required new capital from three countries committed to the tunnel. Although the first link-up, in 1880, was ahead of time, difficult Alpine geology meant that it took nearly two years before the tunnel was open for traffic. Favre, however, had lost not only his fortune but his life, having died of exhaustion in 1879 – a fate shared by many of the labourers he had employed to work on the tunnel.[9]

The southern end of the Simplon, the second of the two great Swiss Alpine tunnels, was actually in Italy, but otherwise the problems involved in its construction were similar to those that had arisen with the Gotthard. Particularly on the Swiss side, however, the route leading to the tunnel mouth was much easier, since for most of its distance it had the benefit of the easy gradients of the wide valley of the upper Rhône; the approach on the Italian side was more difficult, but even then only one spiral tunnel was needed. The real problems came with the tunnel itself, which had to be bored through very poor rock. The work took seven years, so that the tunnel was only open – and that for single-line working – in 1902. A tunnel for a second track had to wait until 1921.

American Railway Imperialism

Turning from Europe to the US leads to a remarkable chapter in the history of railways. The story opens in 1846, when a dispute about territory claimed by both the US and Mexico led to a bitter two-year war between the two countries. The war, decisively won by the United States, ended in 1848 with the Treaty of Guadalupe Hidalgo, which granted to the United States (for $15,000,000) a vast territory, including the whole of California – first explored by Spain in the sixteenth century. The treaty also established the present frontier between the United States and Mexico, together with the 42nd parallel (north) as the northern frontier of California.

Following the end of the Mexican War California (which joined the Union as the 31st state in 1850) developed very rapidly. The discovery of gold in 1849 attracted such vast numbers of prospectors that the California Gold Rush came to epitomize the whole character of this new part of the United States. With a whole new infrastructure required to serve the miners, together with a vast number of hangers-on, the impetus to the local economy was incalculable. In particular, the proximity of the mines to San Francisco (which the American Navy had captured from the Spaniards in 1846) encouraged the development of the finest natural harbour on the Pacific coast, together with new shipyards across the bay opened in 1854. Communications overland were still the great problem so long as they were confined to the stagecoach, the pony express and, from 1861, the electric telegraph. With statehood came an immediate demand for railroads, and in 1853 Congress authorized the survey of a route linking California with the east. This was seen stateside as the only solution, for as one San Francisco newspaper noted in 1856, 'the railroad is of more value to us than the election of forty Presidents'. In 1854 T.D. Judah had already opened a local railroad from the state capital, Sacramento, to Folsom – of great use to the gold miners – but this was never seen as the beginning of a transcontinental line. This only became possible after Judah, in 1860, had persuaded four local entrepreneurs, Leland Stanford, Collis Huntington, Mark Hopkins and Charles Crocker, to meet in a Sacramento hardware store in 1860, and organize the finance necessary for a new railroad at a level that would pay for the final section in California, inland from San Francisco, which would not qualify for federal government support. Such were the origins of the Central Pacific Railroad. The role of Stanford proved to be critical when, as the first Republican governor of the state (1861–63) he secured both substantial investment out of public funds and land grants along the proposed route. The fact that he also ensured that California would be allied with the North in the Civil War in 1862 opened the way for the US Congress – without any members representing the Southern states – to pass the Pacific Railroad Act, enabling the construction of the first transcontinental railroad. This task was entrusted to two companies: Stanford's Central Pacific building eastward from California, and the new Union Pacific westward up the valley of the Platte river – a tributary of the Missouri – from Omaha,

Nebraska. The Central Pacific, benefiting from generous federal, state and county subsidies started construction almost immediately, so that by 1865 revenue from track already built covered costs. There was, however, an acute labour shortage because men working on the line continually deserted for the gold mines; from 1865 onwards this problem was solved by recruiting Chinese immigrants, with the number employed eventually reaching 10,000. The race was on to link up with the Union Pacific approaching from the east with Irish labour.

At a ceremony at Promontory Summit in the Utah territory just north of the Great Salt Lake, on 10 May 1869, with Leland Stanford driving a golden spike, the two railroads finally linked up – an event communicated by telegraph across the nation. 'Jupiter', the locomotive of the Central Pacific train, burnt wood, while Locomotive No. 119 of the Union Pacific burnt coal. The actual link-up was foreshadowed in section 10 of the Act of 1862:

> The Central Pacific Railroad Company of California after completing its road across said State, is authorized to continue the construction of said railroad and telegraph through the Territories of the United States to the Missouri River ... upon the terms and conditions provided in this act in relation to the Union Pacific Railroad Company, until said roads shall meet and connect.

This should be read together with section 11:

> That for three hundred miles of said road most mountainous and difficult of construction, to wit: one hundred and fifty miles westerly from the eastern base of the Rocky Mountains, and one hundred and fifty miles eastwardly from the western base of the Sierra Nevada mountains ... the

> bonds to be issued to aid in the construction thereof shall be treble the number per mile hereinbefore provided ... and between the sections last named of one hundred and fifty miles each, the bonds to be issued to aid in the construction thereof shall be double the number per mile first mentioned.

One does not have to be very astute in reading between the lines to realize that the construction of some 300 miles of track was fraught with difficulty. Over a distance far greater than that to be covered by the new European railways across the Alps, appalling conditions, in a terrain of mountain and desert, made working on the track extremely dangerous, particularly for the coolies employed by the Central Pacific. The Chinese had first come to California in 1854 to work in the gold mines, where they soon acquired a reputation for accepting difficult and dangerous tasks scorned by other labour. Although their work on the railroad was even more arduous, they kept on coming. Their contribution was so important that in 1868 the US agreed the Burlingame Treaty providing for unrestricted Chinese immigration. In 1882, the number of Chinese immigrants peaked at 39,579; by this time, however, their presence was so unwelcome to other West Coast residents that in the same year the US Congress passed the Chinese Exclusion Act which, until its repeal in 1943, effectively excluded almost all immigrants from China. Even so, the Chinese coolies' contribution to the opening up of the West Coast was immeasurable: without them the transcontinental railroads would never have been built.

The Union Pacific was only the first of the transcontinental railroads, so much so that during the 1870s the US almost doubled the aggregate length of its rail network to the point that it comprised more than 100,000 miles of track. During this period, some 51,000 miles of new lines – equal to the entire new construction in the rest of the world – were opened, mainly west of the Mississippi.[10] The Southern Pacific reached Los Angeles in 1876, followed by the Santa Fe Railroad in 1885. In the north, the completion of the Northern Pacific, which reached Tacoma in 1887, opened the way for Washington to become the 42nd state of the Union in 1889. By this time the Canadian Pacific Railroad, completed in 1885, had reached Vancouver. By 1890 the Southern Pacific Railroad also linked Los Angeles with Portland, Oregon, so that almost the whole Pacific region had its own network, independent of the transcontinental lines.

While all these railroads were being built a new factor began to affect their profitability. In the planning and early construction stage, during the 1860s, their promoters had reckoned on transcontinental traffic from the East Coast ports; in 1869, however, the opening of the Suez Canal – just six months after the American railroads linked up at Promontory Summit – to provide a new sea route from Europe meant that much of this traffic was lost. The winners were the West Coast ports, which in the following half century would develop very rapidly.

Canada also acquired its own transcontinental railway. After the British Parliament granted independence to the dominion in 1867, British Columbia, with the entire Pacific coastline, together with countless off-shore islands – of which the largest, Vancouver Island, provided the site of the capital city, Victoria – joined Canada as its sixth province in 1871, but only after it had been promised a railroad joining it to the rest of the country. Even so, it was only in 1885 that the Canadian Pacific Railroad's famous line across Kicking Horse Pass in the Rockies reached Vancouver – so making it the only railroad company whose lines crossed the American continent.[11]

The coming of the railroads had critical economic and demographic consequences for the North American West Coast. During a first phase of actual construction there was always work for the coolies and other unskilled labour, but with the gradual transition to the second phase, which started locally on the completion of every new link, the position changed radically. Low-cost labour came to be seen as a threat to the well-being of the new settlers, which explains the Chinese Exclusion Act of 1882. The new railroads also made the price of eastern manufactured goods highly competitive in West Coast markets – while at the same time, except in British Columbia, the local natural resources included neither coal nor iron for industry. Local agricultural produce also suffered from eastern competition, particularly after the introduction of refrigerated railroad cars during the 1880s. The more railroads were built, the more they competed on rates, making local production even more uncompetitive. At the same time lower passenger rates brought a new wave of overland immigration, leading to a real estate boom and high-cost agriculture. The result was that both manufacturing industry and agriculture, to survive on the West Coast, had to find a new direction: alternatively, the local economies had to find new natural resources to exploit, or West Coast commerce, new outlets. The new railroads fought price wars to attract freight; the question was, what form should it take? There proved to be a variety of answers, but one in particular shows how the west, helped by the new railroads, learned how to exploit its unique resources.

In 1891 Frederick Weyerhaeuser, after successfully exploiting lumber in the Mississippi valley, set up shop in St Paul, Minnesota, where he knew James J. Hill – an entrepreneur with a long track record of success – whose Great Northern Railway, linking St Paul with Puget Sound on the West Coast, was nearing completion. This railroad was a considerable gamble on the part of Hill since the Northern Pacific had already opened with its Pacific terminus at Tacoma. Hill, however, was convinced he had a better route; what is more he had acquired along the line of rail vast tracts of timberland which he sold to Weyerhaeuser at knock-down prices, foreseeing that lumber harvested by Weyerhaeuser would provide the Great Northern's main freight to the east. The immediate result was that Weyerhaeuser's most valuable assets were close to the West Coast; in 1900, after the purchase of 900,000 acres of north-west timberland this led to the establishment of Weyerhaeuser Timber Company – which in the course of time would become the world's largest timber business – at Tacoma, Washington.

With J.J. Hill's Great Northern finally reaching Seattle in 1893, his standing in the Pacific north-west came to rival that of Huntington and Stanford in California. He was at least as enterprising in finding new traffic for his railroad. While timber-loaded Great Northern freight cars travelled on their journey east, demand from the West Coast for goods from the eastern states was insufficient to load them on the return journey. Hill therefore invested in the port of Seattle, while at the same time finding markets in China and Japan for cotton and cotton goods from the eastern states, which then provided west-bound freight for his railroad. This was also critical for the development of Seattle as a major West Coast port – a process encouraged, after 1897, by gold strikes in Alaska and the Yukon.

In California the cheap freight rates forced local farmers to concentrate on produce that could compete in national markets. In California land that had seemed unpromising was transformed by irrigation, which made it ideal for vineyards, first planted in 1861, and later, citrus fruit, with the first trainload of oranges shipped from Los Angeles in 1886. Surprisingly perhaps, dairy products came to contribute the largest share of farm income, even though the state also led, nationwide, in the production of wine – for which it has long had a world market.

In 1882 the four men who had met in the Sacramento hardware store in 1860, by acquiring control of the Southern Pacific Railroad, which linked Los Angeles with New Orleans, sealed their domination of California's railroad economy, the more so since they had also bought out both river and sea steamship lines. While Federal land grants had made the railroads the largest landowners in California, they did not hesitate to use discriminatory freight rates to make or break anyone in mining, manufacture, agriculture or commerce, using this power as a lever to corrupt politics.

By the end of the nineteenth century the four men from the hardware store had been eclipsed by one single man, E.H. Harriman (1848–1909), who had appeared from the east to beat them all at their own game. The son of an Episcopal minister in upstate New York, Harriman, having started as an errand boy on Wall Street in 1862, became a member of the New York Stock Exchange in 1870, and by the 1880s he was busy buying up railroad stock – making a speciality, in the early days, of buying up bankrupt railroads. In 1897, just short of fifty, he became a director of the Union Pacific Railroad; in 1898 he became chairman of the executive committee, and in 1903 president of the company, but by then he was also – since 1901 – president of the Southern Pacific Railroad, so that in the opening decade of the twentieth century his word was law on the consolidated Union/Southern Pacific system, which his own vision had created.

Harriman, ever on the lookout for commercial advantage, also became involved in the railroad politics of the Far East.

In 1905, after war between Japan and Russia had ended with the Treaty of Portsmouth – orchestrated by US President Theodore Roosevelt – a very strong American delegation, led by Secretary of War (and later Roosevelt's successor as President) William Howard Taft, arrived in Tōkyō to settle future relations between the US and Japan. The essence of the agreement reached between the two countries was that the US would allow Japan a free hand in Korea, while Japan would keep its hands off the Philippines – lost to the US by Spain as a result of its defeat in the Spanish-American War of 1898. E.H. Harriman arrived a week later than Taft, with his sights set on the Korean railways, which, as a result of the successful completion, by Russia, of the Trans-Siberian Railway in 1904, could be the key to considerable new traffic. After all he had also acquired, next to all his railroads, the Pacific Mail Steamship Company. However, as a result of what had been agreed by Taft, Harriman had to deal with the Japanese Prime Minister, Count Katsura, who closed the door on him in Korea. The treaty of Portsmouth was extremely unpopular in Japan, which saw in it a US strategy to deny Japan the spoils of its recent victory over Russia. Korea was the most important prize from the wars Japan had fought in the period 1895–1905. The last thing any Japanese Prime Minister could countenance was selling its railways to an American businessman. For once Harriman was defeated.

At his death in 1909 Harriman controlled not only most of the railroads serving California but also two main eastern railroads, the Illinois Central and the Central of Georgia, not to mention other businesses such as his steamship company and the Wells Fargo Express Company. His estate, worth anything from $200 million to $600 million, was left entirely to his wife. In retrospect, the success of Harriman does not detract from that of his forerunners in railroads to California. Leland

Stanford's legacy, today, is a noted university – around which the whole electronics industry of Silicon Valley was built up – and Collis Huntington's is an equally noted art gallery.

In Los Angeles, California, by the end of the nineteenth century, had acquired a city that would rival San Francisco. The two cities were linked by the Southern Pacific Railroad in 1876, but it was only in the 1880s that Los Angeles acquired direct links to the eastern states – of which that provided by the Santa Fe Railroad,[12] opened in 1885, proved to be the most important. The balance of the West Coast economy was soon to change decisively in favour of southern California, and that for a number of reasons. First and foremost were oil strikes, both on land and offshore, in the Los Angeles area. Not only did this lead the western railroads, followed by shipping, to convert to oil – in place of coal from the eastern states – but it also provided them with valuable new traffic. With production only starting in 1892, the yield by 1895 had climbed to 700,000 barrels a day. Then in 1897, Los Angeles was chosen by the Federal Government as a site for special development as a major port with its own shipyards. At a time when the railroads – notably E.H. Harriman's Southern Pacific – as the result of aggressive lobbying effectively controlled, in their own interests, the ports of San Francisco and the bay area, the people of Los Angeles were demanding a municipal harbour. Against strong railroad opposition they got their way in 1907 when the City Council created the Los Angeles Board of Harbor Commissioners – at a time when San Francisco was still recovering from the devastation caused by the great earthquake and fire of 18 April 1906. By this time four shipyards employed some 20,000 workers, while thousands of others worked in some dozen canneries on tuna caught by the local fishing fleet. Los Angeles was ahead of San Francisco, and intended to remain so.

Next to the people of Los Angeles, Harriman's only effective opponent was J.J. Hill, operating at the far northern end of the US West Coast. Hill's attempt to extend his Great Northern Railroad south of the Columbia river into Oregon was bitterly fought by Harriman, who threatened to extend his Southern Pacific north of the river into Washington. They finally compromised by agreeing to the joint use of the Northern Pacific tracks between Portland and Seattle. It should be noted – as something worth more than just a footnote – that by this stage railway workers were also beginning to organize themselves. In 1901 they founded in Oregon the United Brotherhood of Railway Employees, which arranged its first successful strike in Vancouver in 1903.

All this was the realization of a prophecy made some forty years earlier by Henry George, a California journalist: 'The locomotive is a great centralizer. It kills little towns but builds up great cities, and in the same way kills little businesses and builds up great ones.'[13] This proved to be only too true. More than in any other part of North America, the West Coast is dominated by big cities: six of them, San Diego, Los Angeles, San Francisco, Portland, Seattle and Vancouver all count among the thirty largest of Canada and the US combined.

10

ASIA, AFRICA AND LATIN AMERICA: RAILWAYS AND IMPERIALISM

The sixty odd years from the middle of the nineteenth century until the outbreak of World War I in 1914 witnessed the construction of most of the world's railways outside Europe and North America. Since, with the exception of Japan, Korea and China in Asia and of almost the whole of Latin America, this part of the world was divided up in the second half of the nineteenth century between European imperial powers – largely on the terms agreed between them in 1878 at the Congress of Berlin – railway construction was the result of European initiative and investment. The same was largely true of China, Japan and Latin America. In the race to build railways Britain was the indisputable leader, with British enterprise and capital accounting not only for five-sixths of all African and half of all Asian railways,[1] but also for those in Latin America. Particularly in Argentina, where a vast network developed for the transport of agricultural exports, led by beef products, British economic imperialism was the name of the game – a matter about which populist leaders left local populations in no doubt at all, as they did also in China on the other side of the world, where France, Germany, Russia, Japan and the US were also tarred with the same brush.

A comprehensive survey of all the different cases, in Asia, Africa and Latin America, would add up to historical over-kill. First, however, it is worth asking what lay behind such massive investment in railways. India provides the best answer to this question: there, in the years under British rule, the world's largest network of railways – at least in terms of the numbers employed – was constructed, a process which started with the opening of the 21-mile-long line between Bombay and Thane in 1853. India, then, is the first case to look at.

Indian Railways

Although the East India Company, the commercial enterprise that governed India in the first half of the nineteenth century, considered the construction of railways in the 1840s – with considerable encouragement from the new steamship lines operating from Britain – it was very slow in getting its act together. Even at such an early stage Indian government policy was formulated in terms of a comprehensive network – mainly as a result of the vision of Lord Dalhousie, Governor-General from 1847 to 1856, who appointed a consulting engineer for railways for the years 1850–1. This led to a standard gauge of 5 foot 6 inches being decreed for the whole of India, and on this basis railway companies were promoted with names such as the Great Indian Peninsular Railway (GIPR) – calculated to foreshadow future expansion on a grand scale – to build lines in different parts of the country.

As in other parts of the world, the first lines were built to link, with their hinterland, the great ports, such as Bombay, Calcutta and Madras, operated by the East India Company. When the GIPR became the first to open its line, from Bombay to Thane, for some reason the Governor of Bombay, Lord Falkland, chose to leave for the hills the day before the opening ceremony on 16 April 1853, but his successor, Lord Elphinstone, was present a year later, on 1 May 1854, to preside over the celebrations on the opening day of the extension of the line from Thane to Kalyan. This was in fact a more remarkable achievement: first a bridge with twenty-two stone arches had to be built across the Thane creek, and then two tunnels through the hills between the creek and Kalyan.

The civil engineering required over a distance of only 10 miles gave some idea of the obstacles to be expected in constructing railways in India. There were also many possible hazards not encountered in the 30 miles of track between Bombay and Kalyan. How could railways be built so as to withstand the ravages of the monsoon floods, which every year left large parts of the country under water? Would storms blow trains off the track? Would termites eat the sleepers? And most important of all, how would the people of India take to the possibility of long-distance travel brought by the railways?

The celebrations at Kalyan in 1854 gave some idea as to how this last question would be answered. According to a contemporary report: 'Thousands upon thousands came to see that wonder of wonders. The whistle of the engine as it dashed on its glorious course was thought to be the voice of the demon.'[2] This echoed another report relating to the opening of the first section of the railway in 1853, describing the event as:

> a triumph, to which, in comparison, all our victories in the East seem tame and commonplace. The opening of the Great Indian Peninsular Railway will be remembered by the natives of India when the battlefields of Plassey, Assaye, Meanee and Goojerat have become the landmarks of history.

Subsequent history has largely justified such grandiose pretensions: if steam engines were still seen as demons, the trains behind them were seldom short of passengers from every level of Indian society.

In 1855, on the other side of India from Bombay, the East Indian Railway (EIR) was opened from Howrah to Raneegunge, some 30 miles away. Economically this was a very important link: Howrah was on the opposite side of the Hooghly River from Calcutta – the most important harbour in Bengal – and Raneegunge had coal mines, a key asset given the great number of steamships calling in at Calcutta. The opening of the EIR was inauspicious: the ship bringing the locomotives from England was diverted inadvertently to Australia while that with the carriages sank entering the Hooghly river. The locomotives made it to Calcutta in the end, while John Hodgson, the EIR's Locomotive Chief Engineer – showing the resourcefulness expected from mid-Victorian empire builders – designed new carriages to be built locally. Howrah Station, little more than a shed in 1855, was rebuilt in red brick a half-century later, in 1906, to be the largest passenger terminus in India: by then a great deal of water had flowed under the bridge.

On 1 July 1856 the first 63 miles of the Madras Railway, linking the third of India's great seaports to Arcot inland, was opened. On this line, as on the GIPR and the EIR, traffic returns were well up to expectations. However favourable the investment climate for new railways – which sooner or later would link up to constitute a comprehensive system for the whole of India – the political climate was rapidly deteriorating. The Great Mutiny, which broke out at Meerut, some 30 miles north of Delhi, in May 1857, put an end to new construction; indeed, the Mutiny spread rapidly down the route of the new line planned to link Delhi with Calcutta, while the British used the sections of it already opened – such as that between Allahabad and Cawnpore – for transporting loyal troops to fight the mutineers. Where they could, the latter pulled up the track; at the same time railway personnel, led by expatriate British, played a heroic part in suppressing the Mutiny. Although this process was largely complete by the end of 1857, the East India Company did not survive it: the British government took over, to administer India from the India Office in London, with its rule enforced by the Viceroy in India. The new government never doubted the strategic importance of railways: many thought that if the line to Delhi had been completed five years earlier there would have been no Mutiny. However that may be, as every new line opened, the transport of troops and war material was almost invariably part of its operations.

Construction was slow in the years immediately following the Mutiny, so that in 1860 the total length of track was only 838 miles. As Map 6 (page 216), relating to 1861, shows, the networks reaching out from Bombay, Calcutta and Madras – although considerably longer with, critically, the entire line between Calcutta and Delhi open – were still quite separate both from each other, and from two lines in what is now Pakistan. Twenty years later in 1881, by which time the length of rail had increased six-fold to nearly 10,000 miles, railways – as Map 6 shows – covered substantially the whole of India. They did not, however, constitute a single uniform network: this was the result of a decision of the Indian government, in 1869, to build state railways to supplement those of the existing companies. A year later, in 1870, the Viceroy, Lord Mayo, confronted by a shortage of funds while being at the same time concerned to convert India to the metric system, decided to kill two birds with one stone by prescribing a metre gauge for the new lines. Putting an end to Lord Dalhousie's standard gauge had disastrous consequences: the government's policy was interpreted as a licence for building new railways to any gauge that suited local needs – so that, during the 1870s, for instance, the principality of Baroda constructed an extensive network with a 2 foot 6 inch gauge.[3] Not surprisingly, as long as the British ruled India metrication got nowhere. This came with Indian independence in 1947, when the new republic of India also set about converting the railways to Lord Dalhousie's standard gauge. Cases – such as that of the 2-foot-gauge Darjeeling-Himalayan Railway – where local topography ruled out any such conversion – were exceptional.

Returning to the final twenty years of the nineteenth century, one can see the pace of new railway construction slowing, while the mileage still more than doubled – as Map 6 shows – largely as a result of new lines built in the heart of India, as defined by the flood plain of the Ganges and its tributaries. In this process India was consistently far ahead of the rest of what is now called the 'developing world'. In 1870 its railways comprised 55 per cent of the combined operating mileage of all the railways in Africa, Latin America and Asia outside Russia.[4] If in that year there were as yet no railways in either China or Japan, even so, in 1900 – when both countries had their own systems – India still accounted for some 65 per cent of all the railway mileage in Asia. Growth continued in the twentieth century, so that when India and Pakistan became independent states in 1947, the combined length of their railways was 44,722 miles. This was in fact little more than twice the total length of 19,585 miles of British railways in 1923, which served a country with only a tenth of the population and one seventeenth of the area of India. The question for the historian is simply: what did all this add up to for India?

British imperial India was a vast land of extreme diversity in almost every possible aspect, whether defined by geography, economics, demography, culture or religion. In spite of Lord Mayo, its railways were, par excellence, an over-arching institution created by the British to reach almost every corner of the country (as shown by Map 6); what is more, having survived and flourished after the end of British government in 1947, the railways, with their uniform standards of practice, still serve to unify post-imperial India – and to a lesser extent Pakistan and Bangladesh. If the same could be said of the armed forces of these three states – another essential part of the British legacy – in the everyday life of their civilian populations the railways are unrivalled as a unifying force.

When Indian railway construction started in the 1850s, there was, as in the rest of the world, almost nothing of local origin to build on. The British, ahead by a generation in which – by leading the world – they had established workable standards of construction and operation, effectively offered a complete package, as they did also to many other countries, such as notably Argentina – whose railway history is related later in this chapter. The question was what sort of package would be best for India, accepting always that British interests – mainly economic and so governed by the principle that 'he who pays the piper calls the tune' – would be paramount. None the less the process of adapting these interests to local circumstances created a distinctive and original railway world.

The geophysical challenge, in many different shapes and forms, was on a scale unknown in Europe. Mountains, rivers

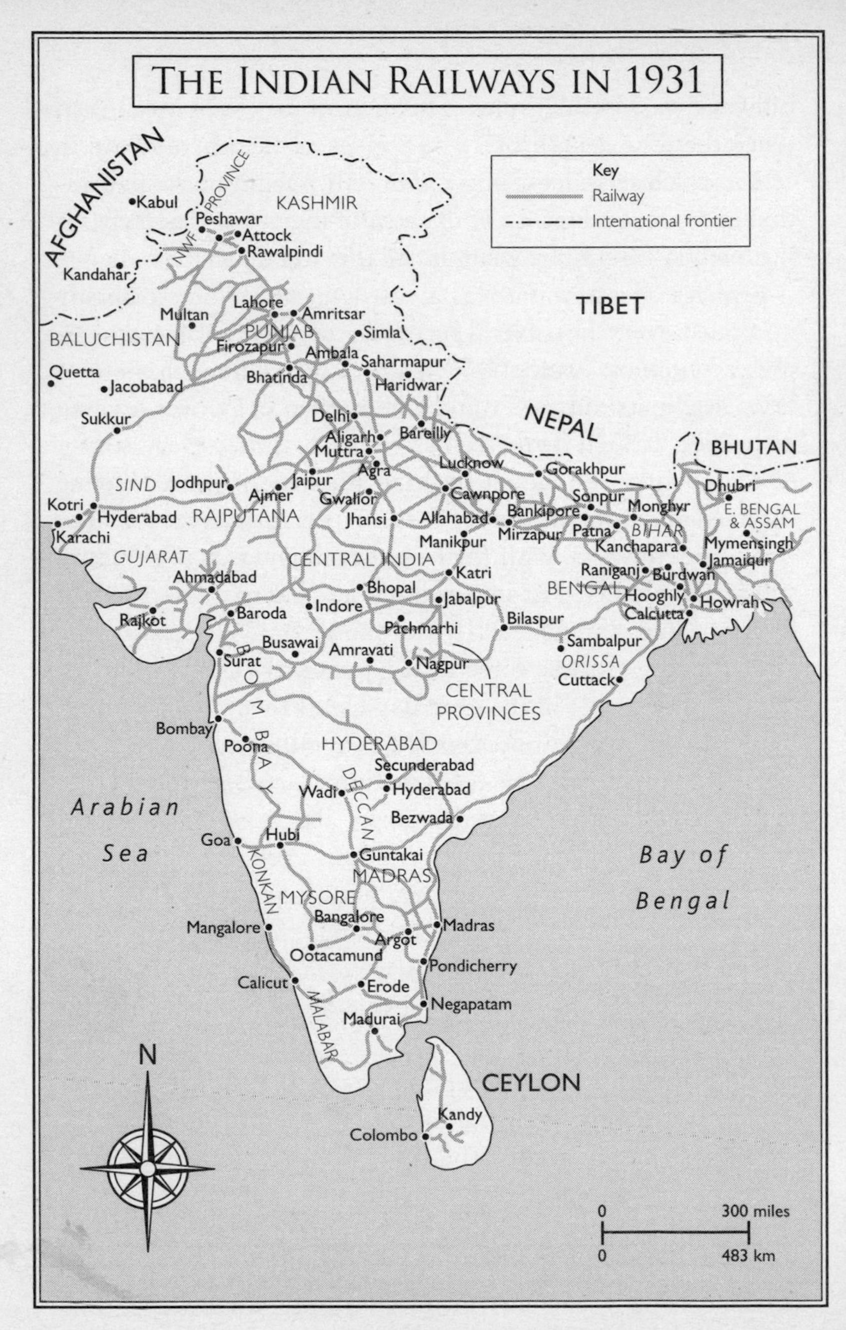
THE INDIAN RAILWAYS IN 1931
Key
Railway
International frontier
AFGHANISTAN
Kabul
Kandahar
NWF PROVINCE
Peshawar
Attock
Rawalpindi
KASHMIR
TIBET
Multan
Lahore
Amritsar
PUNJAB
Simla
BALUCHISTAN
Firozapur
Ambala
Saharmapur
Quetta
Jacobabad
Bhatinda
Haridwar
Sukkur
Delhi
NEPAL
Bareilly
Aligarh
Muttra
BHUTAN
Agra
Lucknow
Gorakhpur
SIND
Jodhpur
Jaipur
Ajmer
Gwalior
Cawnpore
Sonpur
Dhubri
Kotri
Hyderabad
RAJPUTANA
Jhansi
Allahabad
Bankipore
Monghyr
E. BENGAL & ASSAM
Karachi
Mirzapur
Patna
BIHAR
Manikpur
Kanchapara
Mymensingh
GUJARAT
CENTRAL INDIA
Jamaiqur
Raniganj
Burdwan
Katri
Ahmadabad
Bhopal
BENGAL
Jabalpur
Hooghly
Howrah
Indore
Calcutta
Rajkot
Baroda
Bilaspur
Pachmarhi
Sambalpur
Busawai
Amravati
ORISSA
Surat
Nagpur
Cuttack
BOMBAY
CENTRAL PROVINCES
Bombay
Poona
HYDERABAD
Secunderabad
DECCAN
Wadi
Hyderabad
Arabian Sea
Bezwada
Goa
Hubi
KONKAN
Guntakai
Bay of Bengal
MADRAS
MYSORE
Bangalore
Mangalore
Madras
Argot
Ootacamund
Pondicherry
Calicut
Erode
MALABAR
Negapatam
Madurai
N
CEYLON
Kandy
Colombo
0
300 miles
0
483 km

Map 6

and the monsoon rains, presented new problems to the civil engineers. While the future prosperity of Bombay, as a port, depended on extending the GIPR far inland beyond Kalyan – to destinations such as Poona and Delhi – this required the railway to climb the Western Ghats, a long and spectacular ridge just inland from the coast: on the line to Poona, by way of the Bhore Ghat, trains required special banking engines to enable them to climb the steep inclines, while at one point they had to reverse at a special station built only for this purpose. The civil engineers also had to build twenty-five tunnels and twenty-two bridges. This early line was but a foretaste of the problems that would be encountered later in the century, when railways were built in other parts of India – such as the North-West Frontier Province at the foot of the Himalayas.

The flood plains of the great rivers, the Ganges and the Brahmaputra, also presented formidable engineering problems, particularly in the vast, densely populated delta area where they join to reach the sea in the Bay of Bengal. There, as related by a traveller at the end of the nineteenth century:

> the drainage problems to be encountered and solved by the engineers of the railway were of exceptional and extraordinary magnitude. Over the wide expanse of level country, subject to an excessive tropical rainfall, inundations from the flood spill of the enormous channel of the Ganga and other great rivers are often spread as a vast sheet over miles of country, converting the whole district into a semblance of an inland sea from which only the inhabited villages along the higher marginal levels emerge.[5]

It was not enough to build bridges and causeways, simply because the soil, consisting to a depth of several feet of silt brought down by the rivers, would not support them. If bridges were not to be washed away by rivers in flood from the monsoon rains, they had to be supported on piers reaching down to a firm sub-stratum – which could be as much as 140 feet deep. For every single pier this meant sinking a well during the dry season, which was then bricked in to support the part of the pier projecting above the flood water level to carry the iron girders of the bridge itself – which as often as not were imported from Britain. Even then, if a pier was not to be toppled over after eddies from the river in flood had scoured the silt surrounding it, it was best to dump large quantities of stone and rubble on the upstream side.

Much more could be said of the civil engineering problems, but whatever solution was found for any particular case, it was certain to involve the massive employment of unskilled labour in conditions at least as harsh as those endured by such other mid-nineteenth-century railway labourers as Irish navvies in Britain, Chinese coolies in America or serfs and convicts in Russia. The fact that disease and accidents took the lives of about a third of the workforce used on the line across the Bhore Ghat – which on average numbered 30–40,000 people – gives some idea of the working conditions. Even so there was never a shortage of labour: the prospect of paid employment attracted men from far and wide.

The same was true of the prospect of actually travelling by train, even though conditions in the third and fourth classes were appalling – particularly in hot weather. Whatever doubts there may have been in the 1850s about the apparently supernatural powers driving the locomotives, they were soon overcome. For millions of Indians, rich and poor, travelling by train became part of their way of life, so that in 1904, for example, nearly 200 million third-class tickets were sold.[6] The division of carriages into four different classes reflected the class structure of Indian society – although the British who ran the railways consistently refused to take religion and caste into account. The class of any traveller's seat was determined exclusively by how much he was prepared to pay for his ticket.[7]

Indians at every level adapted the railways to their own ordering of society. The Gaekwar of Baroda had a throne installed in the private carriage he used to tour his principality, where his poorer subjects put up with almost any discomfort to travel by train. In 1866, at the other end of India, the British Indian Association of the North Western Provinces, in a petition to the Viceroy, related their concern about the 'dire evil and slavery' imposed on third-class passengers.[8]

The distinctive railway culture extended beyond the trains to the railway stations, where any number of heterogeneous non-travellers added colour to the social scene, which became particularly animated whenever a train arrived. Stops were long, and many of those waiting for trains to come in were there to sell food and drink to the passengers. Others – pickpockets and bag-snatchers – robbed them. With waiting rooms only for first and second class, conditions for other passengers – who often had to put up with trains being hours late – were so bad that according to the 1866 petition to the Viceroy, 'Many a poor Native's illness or death is traceable to sufferings at a Railway Station while waiting for the train.'

Finally, the material culture of the railways, particularly at its most opulent, was astounding. The stations in big cities – not only Bombay, Calcutta and Madras, but many others such as Lucknow and Lahore – were monuments to imperial grandeur, with some, for political reasons, built as fortresses. The largest and most imposing of all was Bombay's Victoria Terminus, named in honour of the Queen Empress, but Calcutta's Howrah Station was hardly less impressive. The imperial connection – which was unmistakable, as it was meant to be – was willingly affirmed by the Indian princes. When, in 1905, the Maharaja Scindia of Gwalior welcomed the Prince and Princess of Wales to his palace,[9] an electric model railway on his dining-room table – with both rolling stock and rails made of silver – circulated cigars and liqueurs.[10]

With independence in 1947, India inherited a chaotic railway situation: in spite of considerable bureaucratic regulation, many of the old companies still operated alongside the state railways, with the independent princely states – such as Baroda and Gwalior, together with many others – retaining their own systems. Little had been done to re-establish Lord Dalhousie's standard gauge. As the princely states were incorporated into the new republic – which without Pakistan and Bangladesh was notably smaller than British India – their railways, together with those of the private companies, became part of one national system, with more than a million employees. The number of passengers was counted in hundreds of millions, with each travelling an average distance which has steadily increased over the intervening years to nearly a hundred miles. The carriage of goods is on a comparable scale. As an imperial legacy nothing else in the modern world compares with India's railways.

Railways of the Andes

Even the mountain railways in India hardly rival those that cross the Andes, in which British enterprise also played a major part. There was only one rationale, which was purely economic: this was to allow the natural resources of the interior, notably nitrates, copper and tin, to be shipped to harbours on the Pacific coast of South America – Antofogasta and Arica in Chile, Rollendo and Callao in Peru. The railways served not only the two coastal states, but also Bolivia, which lost its Pacific provinces to Chile after being defeated in the War of the Pacific (1879–83). The only direct link from the Pacific to the Bolivian railways was from the two Chilean harbours: the Peruvian Southern Railroad linked the seaport of Rollendo with Puno on the shore of the remarkable Lake Titicaca, 12,500 feet above sea level. Traffic for Bolivia then crossed the lake to Guaqui, a distance of some 125 miles, by one of three steamers. (The earliest of these, the *Yavari*, made the long ocean voyage from Hull – where it was built in 1862 – to Rollendo, under its own power. There it was dismantled, to be transported by rail to Puno for reassembly and conversion of its firebox to burn llama dung. The boats that later joined the *Yavari*, of which the largest was the *Inca*, 228 feet long and weighing some 700 tons, also reached Puno the same way, but burnt oil.) Guaqui was linked to La Paz, the Bolivian capital, by some 60 miles of railroad, which in its final descent to the city, from its highest point at El Alto, loses some 1,500 feet of altitude in less than 6 miles. The views of the La Paz valley enjoyed by passengers are, needless to say, stupendous.

The two railways from Chile also met at La Paz; the longer of these, along its route to Antofagasta, passed through Bolivia's main tin-mining region to Uyumi, in the south of Bolivia, which also had a rail link to the Argentine network. Antofagasta in turn was also part of the Chilean railway system whose main line ran parallel to the Pacific coast, to link up with the port of Valparaiso and the capital, Santiago, continuing some hundreds of miles south to Puerto Montz. From Valparaiso there is also a railroad across the Andes to Mendoza, the centre of Argentina's wine production.

With the need to cross mountain passes at heights of more than 13,000 feet, every one of the railway links between the

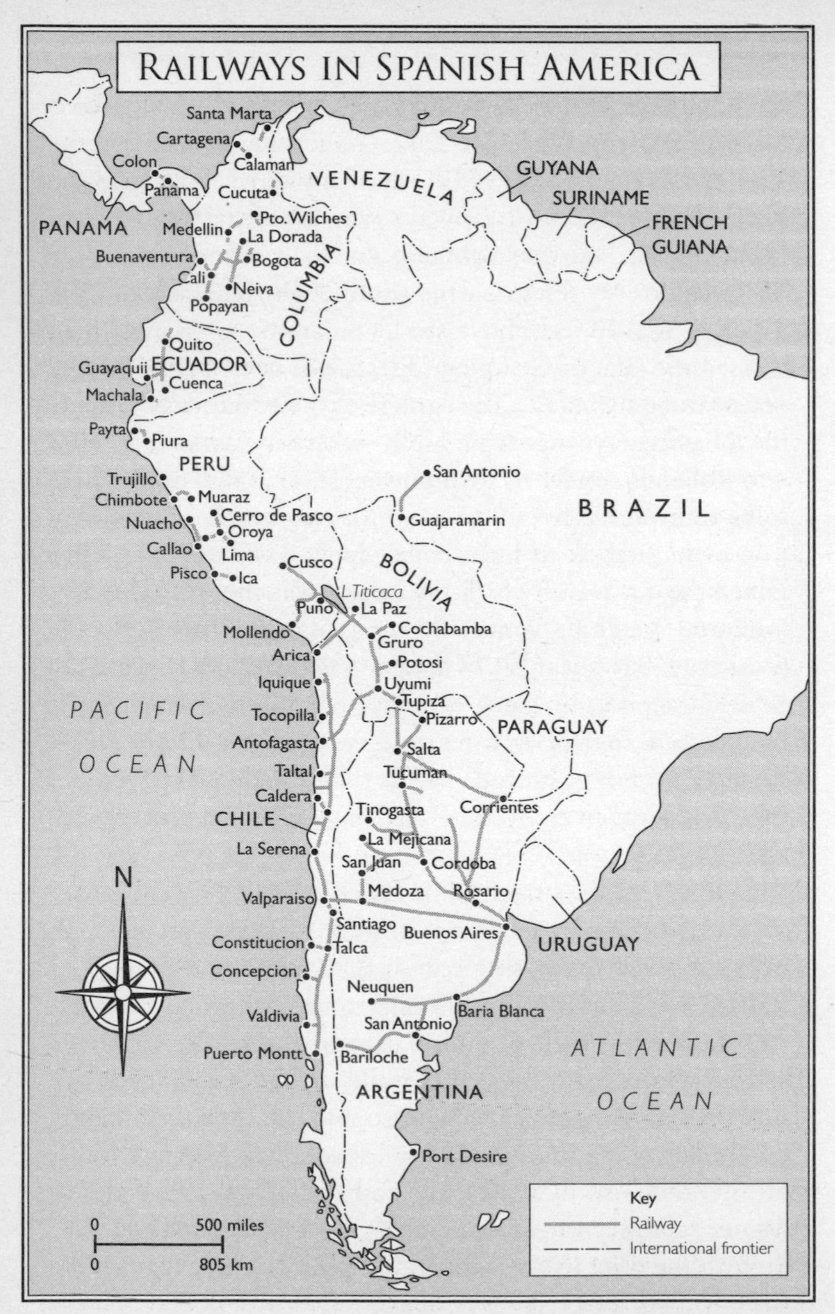
RAILWAYS IN SPANISH AMERICA
Santa Marta
Cartagena
Calaman
Colon
Panama
Cucuta
VENEZUELA
GUYANA
SURINAME
FRENCH
GUIANA
PANAMA
Pto. Wilches
Medellin
La Dorada
Buenaventura
Bogota
Cali
Neiva
Popayan
COLUMBIA
Quito
Guayaquii
ECUADOR
Cuenca
Machala
Payta
Piura
PERU
Trujillo
Chimbote
Muaraz
Cerro de Pasco
Nuacho
Oroya
Callao
Lima
Pisco
Ica
San Antonio
Guajaramarin
BRAZIL
Cusco
BOLIVIA
L. Titicaca
Puno
La Paz
Mollendo
Cochabamba
Gruro
Arica
Potosi
Iquique
Uyumi
Tupiza
PACIFIC
OCEAN
Tocopilla
Pizarro
PARAGUAY
Antofagasta
Salta
Taltal
Tucuman
Caldera
Corrientes
Tinogasta
CHILE
La Mejicana
La Serena
San Juan
Cordoba
N
Medoza
Rosario
Valparaiso
Santiago
Buenos Aires
Constitucion
Talca
URUGUAY
Concepcion
Neuquen
Baria Blanca
Valdivia
San Antonio
ATLANTIC
Puerto Montt
Bariloche
ARGENTINA
OCEAN
Port Desire
Key
Railway
International frontier
0
500 miles
0
805 km

Map 7

Andes and the Pacific coast involved breathtaking civil engineering. The most spectacular of all was not one of the lines to Lake Titicaca, described above, but that of the Peruvian Central Railroad, linking Callao on the Pacific coast, via the capital city of Lima, with Huancayo on the other side of the central cordillera of the Andes – where the rivers are tributaries of the Amazon. After leaving Callao, in just 170 miles the railway reaches it highest point, at 15,700 feet, when it enters the Galera tunnel: of this the last 10,800 feet are gained in less than 18 miles – at an average gradient steeper than 1 in 50. Even with these extremes of gradient a train needs twenty-one zigzags up mountain sides – with the engine alternately pulling and pushing – to gain the necessary height.

I.K. Brunel and Robert Stephenson would have been flabbergasted, and even more by the steepest gradient of 1 in 25: the original proposal for the remarkable Peruvian Central Railroad was made in 1859, the year in which they both died. Over the first 14 kilometres, between Callao and Lima, there was then already a railway, built mainly by convict labour. This, after years of planning – in which British entrepreneurs played a major part – opened as the first railway in Peru in 1850. In 1864 this became the property of Lima Railways, which, as its name suggests, was a British company: it was always popularly known as El Ferrocarril Ingles. It was, however, quite separate from the Peruvian Central Railroad, which also involved British interests, with as main contractor, Henry Meiggs (who would also build the Peruvian Southern Railroad). Work started on 1 January 1870, but by August 1875 only the first 88 miles were complete: further progress was then blocked, first by lack of finance, and then by the four years of the War of the Pacific (1879–1883). The devastation suffered by Peru – which, with Bolivia, lost the war – included the loss of about a third of all its railways. When peace came in 1883 – with considerable territory lost to Chile – Peru was much too poor to invest in its railways. Rescue – if it may be called as such – came in 1889 when Michael Grace, an Englishman, contracted to repair and extend the devastated railway network in exchange for an immediate cash payment, 3,000,000 tons of guano – of which Peru was the world's largest producer – and a 66-year concession to be granted to the Peruvian Corporation, the company he formed in London in 1890. (In the official registration in the Companies Office in London, the concession was stated as being for seventy-seven years. The company did even better when Peru Law 6281 of 1928 provided for all the State railroads to be handed over to the Peruvian Corporation in perpetuity.) The Peruvian Central Railroad finally reached Huancayo in 1908: over the distance of 330 miles from Callao sixty-five bridges and sixty-one tunnels were built, many of them marvels of engineering.

Railways in Argentina

If ever there was a land made for railways, it was Argentina. As stated in one report, sent to London in 1863: 'the ground is almost *naturally* ready to receive rails without preparation.'[11] This was no exaggeration: in every direction overland from the principal city, Buenos Aires, the land – for hundreds of miles – was firm and flat, with few rivers crossing it. At the end of the day, when the railway network was complete, the average construction costs per mile were almost the lowest in the world.[12] (Costs were slightly lower in Australia and South Africa, where the topography was almost equally favourable.) However favourable the Argentine topography, there was still one key problem: nowhere in the vast area of land – centred on Buenos Aires – where the economic advantages of a railway network were most pronounced, were the materials required for its construction to be found. Everything – rails, ballast, wood, locomotives and rolling stock – had to come from outside, as would be true also of coal for fuel. (Only at the end of the nineteenth century, when the Central Norte railway finally reached the forested north of Argentina, did this become the source of wood both for sleepers, telegraph and fence-post, and for burning to make charcoal as a fuel for steam engines.) While such distant provinces might have been able to supply wood for sleepers or stone for ballast, without railways there was no means of transporting them cross-country. The only answer was to open up the country to foreign investors, who could raise the capital needed for construction and provide, at local level, the requisite management and engineering skills. As to the traffic that would justify such investment, this would require the development of an export economy based upon Argentina's vast and largely unrealized potential for agriculture and ranching. Wheat and cattle for the world market, supplemented by sugar, cotton and wine, would – if produced on a sufficient scale – pay for the construction and operation of a comprehensive railway network.

Argentina, after becoming an independent state in 1816, was bound to suffer from its remoteness from Europe – as it had also during the centuries when it had been governed, as part of the Spanish colonial empire, from Madrid. It was only in 1857 that the British Royal Mail Steam Packet Company extended its services to Buenos Aires and the River Plate,[13] and in 1873, that an underwater telegraph cable, laid by the River Plate Telegraph Company, linked Argentina, via Montevideo in Uruguay, to the world-wide cable network. None the less, some among the intellectual and political elite of Buenos Aires were beginning to realize what railways had to offer their country. In 1853, Juan Alberdi, a noted philosopher, stated that 'railways will bring about the unity of the Argentine Republic better than any number of Congresses',[14] realizing that without a great river system there was no other way of populating and developing the interior of the country.

The political context is critical for understanding what lay behind such thinking. For twenty-three years (1829–52) Argentina had been ruled by Juan Manuel de Rozas, a populist dictator, who, with his power base in the provinces, appealed to local patriotism and xenophobia. Anything – such as railways – that might open up the country, was seen as a threat to his regime. In the end de Rozas lost power to an elite based on Buenos Aires, which, in one way or another, would govern the country until well into the twentieth century. Its policy of 'vendepatria' – that is, selling the country – was to exploit Argentina's agricultural potential in the same way as Chile and Peru exploited their mineral wealth. In Argentina, also, railways were just as essential as those already built, in the much more difficult terrain of the Andes, to link the mines of Chile and Peru to harbours on the coast. Although the new governors of Argentina were hardly blind to the fact that everything in Chile and Peru had depended on foreign capital, they themselves took the initiative in the first venture into railways.

Once again politics decided what this should be. Although Argentina, in the 1850s, aspired to be a unitary state, it was then no more than a confederation in which the key province of Buenos Aires – disregarding the interests of the rest of the country – blocked all progress towards a single republic. The result was that the Central Argentine Railway – planned to be the first in the country – was built from Rosario, on the upper Parana river, to the important inland city of Córdoba at the heart of the richest agricultural region. Since the Parana was navigable by sea-going vessels the new railway would open up a vast new agricultural export market. Life, however, was not that simple.

An American engineer, Alan Campbell, was employed to survey the route, which he found – after the Andes where he had earlier worked on railways – a civil engineer's paradise. On his introduction, William Wheelwright – his former employer in the Andes and builder of the first-ever railway in Chile – was granted a share in the project. Even though this soon converted to complete ownership, Wheelwright, in spite of promising access to what would be a new 'Texas' – to say nothing of land grants and exemption from import duties on imported material – failed to attract foreign investment on a scale sufficient to complete the line. So long as Buenos Aires kept out of the Confederation no foreign concern was ready to risk partaking in such a venture.

While Wheelwright was getting nowhere in Rosario, local businessmen in Buenos Aires were busy setting up the Sociedad Anonima Camino Ferrocarril al Oeste, to build a 7-mile line – on the broad gauge of Spain and Portugal – to the western suburb of Flores, where they owned land ripe for development. With a grand opening ceremony planned for 29 August 1857, a group of local dignitaries decided to travel on a trial run the previous evening. On the return journey their carriage derailed and overturned to dump them all in a ditch running alongside the track.[15] Suffering no serious injury, and sworn to secrecy, the dignitaries helped put the train back on the rails, where it successfully completed the journey back to Buenos Aires.

Although the day was rescued, there were still problems about extending the 'Oeste' further out of Buenos Aires. By this time it was clear that the necessary finance must come from British investors. From the very beginning the Oeste's vice-chairman, Daniel Gowland, was also the chairman of the Committee of British Merchants in Buenos Aires, while another Englishman, William Bragg, supervised the building of the line. For this he not only imported its first locomotive, all the rolling stock and tracks from England, but also employed 160 British workmen. Even so, by 1860 only 23 miles, all part of Oeste, had been built in the whole country, so it is not surprising that a 238-mile extension proposed in 1858 met many negative reactions.

The year 1862 saw the end to the difficult political climate with the accession of an enlightened new President, Bartolomé Mitre, who, as Governor of the province of Buenos Aires until that year, had finally persuaded it to join the rest of the country in forming the new Argentine Republic. Knowing that Mitre saw railways as the economic salvation of Argentina, Wheelwright now appealed to him for help in funding the completion of his Central Argentine Railway. On 5 September 1862 this led to a concession with the Argentine Republic guaranteeing investors dividends at 7 per cent for forty years based on construction costs of £6,400 per mile – the standard set by similar concessions granted by India, Canada and Russia – with a land grant extending to 3 miles on both sides of the railway.

At almost the same time the province of Buenos Aires awarded a British entrepreneur, Edward Lumb, a concession for the Great Southern Railway to be built over a distance of 72 miles to Chascomus. The terms were the same as those granted to Wheelwright for the Central Argentine Railway, except that the cost per mile was agreed at the much higher sum of £10,000. Wheelwright, with ambitions to control the whole Argentine network, protested at this discrepancy. This was very unwise, for although the Governor of Buenos Aires delayed in signing the contract with Lumb, he finally did so after an influential friend of Lumb's, the English businessman, David Robertson, had paid out some £22,000 in bribes to local officials. Wheelwright could in no way afford to make an enemy of the British business community in Buenos Aires at a time when London was effectively the only place to raise capital – the US having to use any money that could be raised to finance the Civil War. In the end, in 1865, Wheelwright had to give up his interests in the Central Argentine Railway, which, under his control, had made little progress. With construction delayed as a result of war between Argentina and Paraguay (1865–70), it finally opened in 1870, with a public ceremony in which the Bishop of Córdoba pronounced his official blessing. All this meant in the long run that most railways in Argentina, until well into the twentieth century, would be built and owned by British interests.

What did all this mean for both Argentina and Britain? To begin with, the point is worth making that the relatively late development of Argentine railways, which only got under way in the second half of the nineteenth century, meant that they always operated with state-of-the-art equipment – whether track, locomotives, signal boxes or whatever – as it was at the end of the first generation of British railways. When looking at the balance sheet from an Argentine perspective, its most salient feature is the standard provision for a state-guaranteed minimum dividend. To take the early case of the 72-mile-long Great Southern Railway, the 7 per cent guaranteed dividend, based on the agreed cost of £10,000 per mile, meant a maximum state commitment of £50,400 per annum – but with only a modest level of success this would soon drop to nothing. The guarantee's main virtue was that it gave the state a positive incentive to encourage traffic, and discourage government interference. In the laissez-faire climate encouraged by the line of liberal presidents which started with Mitre (1862–8) this was no real problem. The land-owning elite that governed the country welcomed the railways for the opportunities they provided for profitable speculation in land, claiming, at the same time – with some justification – that the population at large also benefited from much-improved communications, to say nothing of new employment opportunities. In increasing measure, also, the railways helped maintain law and order in distant provinces where local caudillos and unruly Indian tribes were only too ready to challenge Buenos Aires.

Whatever the benefits of a railway system were to Argentina, the real winners were British, whether as members of the local expatriate community – divided between the capital and the great agricultural and ranching estates – or as ordinary citizens of the home country. As to the first category, a substantial expatriate community grew up to exploit the economic opportunities of the Argentine. It enjoyed access to new capital in London coupled with a patrician life style based on Buenos Aires, where the men could play polo at the Hurlingham Club while their wives shopped at Harrods – both clones of the eponymous institutions in London. These pleasures, and the status that went them, were shared with the local Argentine elite, who found them much to their liking. On the other hand they were not much interested in local directorships, preferring to enjoy the enhanced land values that came with every new railway.

Britain itself won all along the line. Limited liability companies, founded according to the Companies Act of 1862, facilitated investment in every new railway project in Argentina – and for that matter the rest of the world. This was increasingly welcome to the capital market at a time when the already mature British railway system provided few such opportunities. What is more, the new overseas railways – particularly when owned and financed from Britain – were a vast new market for British-made locomotives, rolling stock, signal and telegraph equipment. This in turn added enormously to employment opportunities and the size of British towns – which by this time were beginning to enjoy their own suburban rail services. The repeal of the Corn Laws in 1846 – which itself reflected a pronounced shift in the balance of power in Britain – left the way open to the new urban populations to consume agricultural products imported from the other side of the world. For some twenty-odd years, however, the change to imported foodstuffs was a slow process, largely as a result of radical technological improvements making it possible for British farm output to keep pace[16] with the steadily increasing population: when it could no longer do so railways opened up the vast wheat fields of Argentina and western Canada.

Oddly perhaps, the most substantial growth of Argentine railways had to wait until after 1880, by which year there were still only 1,388 miles of track.[17] There were a number of reasons for the slow start. Frontier unrest in Patagonia (which the railways never reached) and the more distant of the great planes of the Pampas consistently destabilized Argentine politics, which only settled down after Buenos Aires became the nation's definitive capital in 1880. At the same time increases in the size and speed of transatlantic steamships, coupled with much lower fares for steerage passengers, brought a flood of immigrants from Italy and Spain, whose presence in the Argentine as agricultural smallholders, shopkeepers and craftsmen, generated considerable new traffic for the railways, where many of them also found employment. There were also significant technological advances: not least among them was the refrigeration of meat, which enormously increased the profitability of ranching – and the demand for barbed wire. All this meant that in the years leading up to the First World War, there was a tenfold increase in the length of track, which in 1915 measured 22,251 miles[18] – a distance that had included, after 1910, a new link across the Andes to Chile and the Pacific Ocean. This was a considerable amount of railway for a total population of only 5.5 million.

If the period 1880–1914 is seen as a golden age in the economic history of Argentina, this did not mean that the British were everywhere as popular as they were among the elite. There were many grievances about how they ran the railways. The goods wagons, based on the standard British single-bogy model, were inconveniently small for the long-distance transport of agricultural products. Only in 1900 did the British-owned railways begin to introduce the standard American cattle car with two four-wheel bogies. While there was too little capacity in the peak season from January to April, the railway owners were reluctant to increase their investment in wagons which for the rest of the year would be left on sidings. At the same time they saw little to gain from providing either mechanical hoists and cranes or decent storage, whether at loading points along the line or at the seaports, which were the final destination of export traffic. Such investment would justify higher rates, when those already charged were generally seen as too high. At the same time there were flagrant abuses of the system of guaranteed dividends, which – as many had foreseen – discouraged best business practice. The worst case was that of the Central Argentine Railway, which had caused so much trouble in the 1860s. Over a three-year period, 1883–5, of exceptionally high profits, the company refused to repay – as it was bound to by the terms of the original concession granted to Wheelwright in 1862 – sums earlier received under the 7 per cent guarantee clause. When the government threatened enforcement of its rights, the railway directors stopped the flow of freight and blocked further investment in extending the system into the northern pampas. The company chairman, Frank Parish, then demanded that the government cancel the debt, and to add insult to injury increased the capitalization by the same amount. He did, however, agree to drop the guarantee, but given the level of profits this was an almost worthless concession. The popular outcry was loud and clear, but led to no action by the government, which feared anything that would jeopardize the flow of foreign capital.[19] The problem for Argentina – and for many other parts of the world outside Europe and North America – was that when it came to capital, Britain enjoyed a seller's market. If Argentina no longer wanted British capital, it could find new opportunities in Brazil, India, Russia or even China.

There was, however, one man in Argentina, who worked hard to check the British directors' abuses of what was close to a monopoly of the country's railway network. A statement made in 1888 by President Miguel Juarez left little doubt about his feelings:

> I cannot find adequate terms to describe the conduct of those companies which collect their guarantee at the end of each quarter and yet pay little attention to traffic which is allowed to languish and decrease. On the slightest pretext they call attention to the necessity of paying the guarantee punctually in order to maintain the credit of the country abroad, but I do not see how the credit of the country could suffer when it is proven that the state is compelled to take coercive steps against companies that have converted government protection into a criminal and iniquitous exaction.[20]

To show that he meant business, Juarez, in October 1888, informed the Argentine Great Western Railway that the guarantee payments would be reduced if it failed to acquire adequate rolling stock. The Company Board, when it learnt of this threat from the Argentine Minister in London, simply brushed it aside, while the British business community in Buenos Aires saw it as no more than an attempt to make the British-owned railways a scapegoat for the financial chaos suffered by the country as a result of Juarez's own reckless spending. They noted, also, that the state-owned railways did no better. In November 1889 Juarez went on to suspend guarantee payments to the East Argentine Railway Company, but then in 1890 the Argentine financial structure collapsed and Juarez was forced out of office. Locally the crisis was the result of borrowing mania – mainly to finance new railways – during the 1880s, but the ruin of Baring Brothers, a London merchant bank, in 1890, and the shock to the whole British financial structure that followed, were the last straw. Under the notoriously corrupt Juarez Argentina lost its entire gold reserves and the local oligarchs the last of their foreign credit.

Vice-President Carlos Pellegrini, who succeeded Juarez, introduced the General Railway Regulation of 1891 as a means of reducing the abuses alleged against the British directors and their supporters' club among the local Argentine oligarchs. It did neither: rather, by allowing favourable rates for large shipments, it made small users pay more. The extension of the Great Southern Railway from the seaport of Bahia Blanca to Neuquén at the foot of the Andes, begun in 1895, showed how little the British companies had to worry about. The new railway was seen as a strategic necessity at a time of apparently imminent war with Chile. In the 'grave national crisis' President Roca (1892–8), having called it a 'new and wonderful testimony of the benefits given to the country by English capital and enterprise',[21] gave the company a 15 million peso government handout, all the land necessary for its right of way – acquired by the government exercising its right of eminent domain – and a forty-year exemption from import duties on all equipment needed for construction and operation: no wonder then that the Great Southern was by 1900 the most prosperous railway company in the country.

The new century brought some reforms: the Mitre (*sic*) Law of 1907 finally allowed the government to fix rates (but only when profits had exceeded 17 per cent for three years), abolish guarantees (in exchange for import duty exemptions similar to those already granted to the Great Southern Railway) and tax railways at a rate of 3 per cent of profits – the proceeds to be used for feeder roads to stations and ports. British companies took advantage of the new law to build 7,700 miles of new line between 1907 and 1914, leaving construction in remote unremunerative areas to the government.

But this was in the end a fools' paradise, and the first signs were already to be seen by 1914. The apparent success of the Argentine economy, and the opportunities it offered to enterprise, attracted ever more immigrants, so that in 1914 some 50 per cent of all residents of Buenos Aires were foreign born: among Spaniards and Italians, followed by Syrians, Lebanese and East European Jews, a rich but alien minority emerged to control 72 per cent of local commerce, while disdained by the local elite – whether British or Argentine. In a parallel process the elite were willing accomplices in the sell-out of their country's railways – the Great Southern, the Central Argentine, the Buenos Aires and the Pacific and the Great Western – to British companies.

In 1912 there was something of a scare – as much in London as in Buenos Aires – when an American syndicate, headed by Perceval Farquhar, who already had a successful track record in other South American states – offered to buy two poorly managed and unremunerative state railways in the north of Argentina. Their sorry state was almost wholly the result of corruption: railway employees at every level held their jobs as rewards for political services, while at the same time tariffs were fixed to suit the residents of districts along the line of the rail. This may have been good for buying votes, but it was 'a hell of a way to run a railroad'.[22] No wonder then that Buenos Aires was ready to sell out to Farquhar. The local English-language press shed a few crocodile tears: 'the "dynamos" operating in New York will have the power behind them necessary to make the countries concerned dance to New York time ... Once the Farquhar trust ... establishes itself firmly in South America, it will drag down the United States on top of all these little Republics'.[23] In the event there was nothing to worry about: setbacks in Brazil and Uruguay meant the end to Farquhar's plans for Argentina. The fact that they represented a considerably better bargain for the Argentine government than anything they were getting from the British did not concern the latter.

There were some, however, who read the writing on the wall at a time when provincial criollo dissidents on one side, and European immigrants – with dangerous ideas about organizing trades unions and even strikes – on the other, were beginning to threaten both the Argentine oligarchy and the British interests which they had so long protected. In 1909, Eduardo Castex, a popular orator, said,

> As our ancestors attained political independence when they had sufficient force to impose it, in the same manner, our successors, at their own time, will attain economic independence, recovering the railways and other public works that are now in the hand of foreign capital when these same works have yielded to the inhabitants of the country the riches necessary for their acquisition.[24]

Although the railway histories recounted in this chapter end with the First World War, it is worth noting, in this case, how the Argentine economy – built up to a high level of prosperity on a foundation of foreign-owned railways – failed in the twentieth century, with disastrous social and political consequences, particularly under the nine-year rule of the dictator Juan Perón (1946–55) and his wife, Evita.[25]

Colonial Railways

The period of some fifty years leading up to the beginning of the First World War in 1914 was the great age not only of steam, but also of overseas imperialism. Although Britain undoubtedly led the way, numerous other countries played much the same game, so that by 1914, seven continental European powers, Belgium, France, Germany, Italy, the Netherlands, Portugal and Spain were all in it, together with Japan and the US. Russia was a special case since its imperialism consisted of extending its own land frontiers. The new empires ruled much of south and south-east Asia, all of Africa except Ethiopia, and the whole of Australasia. South America was a separate case, since Spain and Portugal had lost empires there, while the US – under the Monroe doctrine – guaranteed the independence of the successor states. (The three Guianas, British, Dutch and French, on the Caribbean coast, are a special case, interesting for the fact that the British Demerara–Berbice railway, opened in 1848, was the first railway in South America.) The former Spanish and Portuguese colonies may have constituted an empire under another name, as a result of Yankee imperialism, but when it came to railways, it was mainly British investment that counted – as already related in the sections on Argentina and the Andes. India was a special case, as was China in a rather different sense.

The key fact about the classic colonial railway is that it was built to serve the economic or strategic interests of the imperial power, which also supplied the necessary capital, together with all equipment, including locomotives, wagons and rails, and suitable expatriates to run and maintain the system. Satisfying the interests of local inhabitants, such as for passenger services, was at best a secondary consideration – although it could become very important, as in Australia. The main object was still to exploit either plantations or natural mineral resources, or to help deploy and supply troops – an expensive, but sometimes necessary operation, that Pax Britannica was intended to avoid as much as possible. On the import side the colonial railway would transport European, American or even Japanese consumer goods for the local market: capital goods imported along a colonial railway were mainly such as were necessary for expatriate-owned mines and plantations, and the supporting infrastructure. No imperial power wanted its colonies to compete with its own domestic industry. Indian railways, for example, would ship Lancashire textiles inland, but not imported machinery that would enable textiles to be manufactured locally. (In the long term this was a losing strategy: twentieth-century India, with its vast domestic market, soon had its own flourishing textile industry.) Australia was always a special case, since the colonies were home to British settlers: even so each one of them had its own railway network, with as its hub its major harbour – Sydney for New South Wales, Melbourne for Victoria, Brisbane for Queensland and Adelaide for South Australia. The absence of a uniform gauge inhibited traffic between them, but then, economically, there was little demand for it. The Australian economy is still based on mining and agriculture producing for export – so much so that as late as 1986 Paul Keating, a future prime minister, suggested, in a radio interview, that his country might be a 'banana republic'.[26]

The factors outlined above meant that the typical colonial railway linked a sea-port with an area either rich in mines – such as the high Andes as related on pages 220 to 224 – or devoted to plantation agriculture. The first stage in the transport of coffee, sugar, tea, cocoa or bananas to overseas markets was as often as not along a purpose-built colonial railway. Sumatra, much the largest island in the Dutch East Indies, had three such railways with no links between them. They were of no use to local interests, but that was never their purpose. Only in Java, a smaller and much more densely populated island, did the Dutch build a system useful for local traffic. In the countless other islands of their empire there was not a single railway, and the same was true of many other parts of Europe's colonial empires. In the Philippines, an island state comparable to Indonesia, the Americans' only railway was on the main island of Luzon, running south-east from Manila, the capital city.

Two wars fought by Britain at the end of the nineteenth century – that against the Mahdi in Sudan and that against the Boers in South Africa – illustrate the strategic importance of colonial railways. As to the first, little needs to be added to the passage from Winston Churchill's *My Early Life* quoted on page 113. The 400 miles of railway along which he – and many thousands of soldiers – travelled to confront the Mahdi were laid at the behest of the British Commander-in-Chief, Lord Kitchener. They were essential to the success of the British Army, even though the line stopped some 200 miles short of Omdurman, where the decisive battle was fought. This stretch of rail was the beginning of the Sudan Railways, which went on to become a remarkably comprehensive network described by one author as 'one of the most comfortable systems in the world'.[27]

South African railway history in the decade that ended with the Boer War is more involved. The key factor was the discovery of gold, on a scale far beyond that of California and Australia in mid century, along a low range of hills known as the Witwatersrand – soon known simply as 'the Rand' – located deep inside the free but land-locked Dutch-speaking republic of the Transvaal, whose citizens were mainly farmers – or, in Dutch, 'boers'. The result was a stampede of gold-seekers and fortune-hunters focused on Johannesburg at the heart of the Rand; known to the Boers as 'uitlanders' – that is, foreigners – they were tolerated rather than welcomed. They were certainly not to be granted any rights of citizenship, and yet the whole enterprise was – despite the diverse nationalities of the uitlanders – fundamentally, British, with finance raised in the City of London. At the same time the gold from the Rand had to be shipped to markets overseas, while the equipment required by the mines had to be imported. The Boers, who had only come to the Transvaal in the first place to escape from the British Cape Colony and Natal, had long contemplated a railway link to the sea that would not pass through either of these two colonies. This was entirely realistic, since the closest seaport to the Rand was Lourenço Marques on the Indian Ocean at the southern end of the large Portuguese colony of Mozambique, which had a long frontier with Transvaal; the Netherlands Railway Company, a Dutch enterprise, finally completed the link in 1895. By this time the British had also built railways, from Cape Town and Durban – on the coast of Natal – into the heart of South Africa, but the distance to the Rand was much greater in both cases. What is more, if President Kruger of the Transvaal had anything to do with it, they could get none of the traffic. To add insult to injury, he gave two companies, one French and the other German, a licence for the local manufacture of dynamite – an explosive required in large quantities by gold-mining. The shareholders, few if any of whom were Boers, made a lot of money. The more traditional Boers were not pleased: the most important enterprises in industry and transport were in the hands of foreign shareholders. It could hardly have been otherwise, given the negligible capital resources of the Transvaal Republic – and in any case the British had been kept out.

The events leading up to the Boer War (1899–1902), starting with the Jameson Raid in 1895, are well documented. At the heart of the problem was the exploitation of the greatest mineral wealth in the history of the world by a community of foreigners, with the boom town of Johannesburg as their capital city, living among a population of farmers, so thinly spread across the surrounding countryside, that it was said that if any one of them could see the smoke from his neighbour's hearth, both considered that they were living too close to each other. After 1895, the key issue was the representation of the uitlanders in the Council of the Republic. In the final year of the nineteenth century Kruger had to deal with a very forceful colonial secretary, Joseph Chamberlain, in Whitehall. After months of negotiation both, by September 1899, had concluded that war was inevitable.[28] By this time British troops, massively reinforced during the preceding months, were deployed along the boundaries of Transvaal. Finally, on 9 October the British received an ultimatum containing terms almost impossible to comply with – for both political and logistical reasons – threatening war after forty-eight hours if they were not complied with. The result inevitably was war.

It is time to look at the map, and particularly the way the railways went. When it came to the British forces confronting those of the Transvaal, at the centre of the battlefield, the Boer nation was sandwiched between Natal to the south-east, and the furthest reaches of the Cape Colony to the west; to the south was its somewhat reluctant ally, the Dutch-speaking Orange Free State, with its capital at Bloemfontein. As for railways, Pretoria, the capital, was linked to the sea by the line to Lourenço Marques – which was safely beyond the reach of any British force – and by a short line to Johannesburg, which then continued into Natal and on to the major seaport of Durban – firmly in British hands. This railway was bound to cross the line of battle between the Boers and the British, wherever it happened to be. It was equally bound to be a major supply line, particularly for the British Army. In the west, Kimberley, in the Cape Colony, which had become notorious for the wealth of its diamond mines, and Mafeking, some 250 miles to the north, lay on the railway linking Cape Town to what was then called 'Rhodesia', with a considerable long straight stretch just inside the Bechuanaland Protectorate from its frontier with Transvaal. Cape Town was also linked by rail to Bloemfontein and on to Johannesburg across the Vaal river.

None of these lines had been planned with military traffic in mind, and when the war came, they were hardly up to the demands made on them for the transport of troops and ordnance – so much so that after six months of war Lord Roberts cabled to London for twenty-five engines and 300 wagons as a 'matter of urgency'. It was much too late, but so were the Boers in recognizing that the railways were the weakest link in the British supply chain. A single-line track, laid across mile after mile of open and often featureless countryside, known intimately to countless Boers – most with their own horses – was an obvious target for raiders, particularly where something essential like a bridge could be destroyed.[29] Although sappers were always busy repairing the damage, many a British soldier in the front line went hungry when the trains failed to get through. It is difficult to know whether the British forces, which were constantly separated from each other, suffered more from the inevitable congestion on railways forced to carry traffic for which they were never designed, or from sabotage by Boer raiders.

Turning from the general to the particular, railways played a key role, first in the capture of Winston Churchill by Boer commandos, and second, in his escape from prison in Pretoria. Churchill devotes a whole chapter, entitled 'The Armoured Train', to the events leading to his capture, and two more entitled 'I Escape from the Boers', to his escape. He was captured on the railway from Johannesburg to Durban, along which the Boers had advanced a considerable distance into Natal, scoring one victory after another. He was present not as a soldier, but as a journalist, and as such was entitled to be released after he was captured. When, however, the armoured train on which he was captured was advancing up to the Boer lines in support of British cavalry, and was derailed by enemy action, he played an active role in getting it back onto the line. By so doing, according to the Boers, he forfeited his privileged status, so once they had captured him, they brought him to Pretoria as a normal prisoner of war. To the railway historian the whole illustrates an early instance of the use of trains for military operations – in this case for one which, as Churchill saw at the time, was utterly misconceived.

As for Churchill's escape, he knew from the beginning that travelling by train along the line to Lourenço Marques was the only chance of getting out of the Transvaal. Even so, the risks were enormous, since the entire line up to Komatipoort, just short of the coast – where it entered Mozambique – was in enemy territory. The hue and cry following his escape made it impossible for Churchill, who spoke not a word of Dutch, to travel as a normal passenger. The only alternative was to ride goods trains by night, which, given their slow speed, meant a hazardous and interrupted journey. None the less Churchill's bravado was crowned by success, and once he had announced his presence to the British Consul in Lourenço Marques, the electric telegraph soon brought the news both to Pretoria and London.

11

THE GREAT EASTERN RAILWAYS

Russian Railways

The vast extent of the Russian land mass, extending in latitude from 38° to 76° N and in longitude from 20° E to 170° – which is nearly halfway round the world – always meant that distance represented the greatest challenge to the construction of any line of communication overland. Although the problem of holding together such a vast empire confronted the Russian Tsars as early as the seventeenth century it became much more acute when, in the first half of the nineteenth century, both the US and the major European powers committed themselves to opening up China and Japan to commercial exploitation.

This period of history was also that of the first generation of railway construction in the Western world, as described in Chapter 7. Ever since St Petersburg became the Russian capital in 1712, the country, and in particular the Tsar and the ruling classes supporting him, had to reckon with political, economic and technological developments in the outside world. Tsar Peter I ('the Great') 1682–1725, by bringing in architects, engineers and craftsmen from Western Europe, while at the same time founding – with considerable help from a new German immigrant community – an imperial civil service for implementing government policy, established a tradition of adopting Western institutions to meet the special needs of his empire. Being Russia, there were inevitably reactions, at almost every level of society, from people who – preferring to look inwards at the vast expanse of their own country – resisted innovation from outside.

Such was the climate in the 1830s when railways began to be built in England, the US and any number of different European countries – a development that inevitably led Tsar Nicholas I and his ministers to consider whether they should also be built in Russia. The Tsar himself was in favour, but the two ministers who counted, Kankrin (Finance) and Tol' (Communications), were not so. This became clear in 1834 when Professor von Gerstner, from Vienna – who already had a good track record for building railways for the Austrian empire – submitted a memorandum to the Tsar proposing two routes, one between St Petersburg and Moscow, and the other linking Moscow with Kazan, via Nizhnii Novgorod. In exchange for a twenty-year monopoly von Gerstner promised that his railways would be able to transport 5,000 infantrymen and 500 cavalrymen, together with artillery and horses, between provincial capitals at a speed of 212 kilometres a day. This appealed to the Tsar, who had already noted how the British government had been able to use the much shorter Liverpool & Manchester Railway to transport troops during an Irish emergency.[1]

Kankrin, an economist, advised that the capital required for the proposed railways could better be spent on improving Russian agriculture. He was also concerned that carters would lose trade and that Russia's forests would be depleted by providing fuel for the steam locomotives. To compromise with his ministers Tsar Nicholas approved a short experimental line, 15 miles long, linking St Petersburg with Pavlovsk, a resort on the Gulf of Finland, via Tsarskoe Selo, where he had his own summer palace. Encouraged by very generous financing von Gerstner completed the line to Tsarskoe Selo in 1837, and to Pavlovsk in 1838. With restaurants and a ballroom at Pavlovsk the railway was immediately popular with the capital's inhabitants, particularly since it was committed to set aside an annual sum for free entertainment at Pavlovsk – which in the early 1860s even included concerts conducted by Johann Strauss. All this led Kankrin to observe, somewhat disingenuously, that 'whereas in other countries railways were built to industrial centres, in Russia the first railway led to a tavern ...'[2]

If the line to Pavlovsk demonstrated that Russians could build railways, it little foreshadowed the problems, which would arise with the construction of long-distance lines. Even so, an obligation imposed on von Gerstner, to buy Russian equipment – so long as the extra expense did not exceed the cost of importing via St Petersburg by more than 15 per cent – proved unrealistic: the iron-works in the Urals – to which von Gerstner's second proposed long-distance line would have provided access – were not up to the task. This proved to be a consistent obstacle in the way of Russian railway development. If, in principle, the official policy of exploiting the needs of railways to develop new industry – which Russia sorely needed – was justified by national interest, its implementation, at least in the early stages, left the railways with poor-quality track incapable of carrying the heavy locomotives required for efficient and profitable operations.

Following the success of the line to Pavlovsk, in 1841 the Tsar accepted a very sound and complete proposal – made by a specially appointed commission – for a double-track railway between St Petersburg and Moscow. True to his character, Kankrin's reaction was negative: 'All thinking people abroad consider that it will realize no profit, will ruin morality and liquidate unproductively capital which could be put to better use.'[3] In 1842 Tsar Nicholas set up a committee with his son and heir (the future Alexander II) as chairman to direct the construction of the new railway. Although its original members included both Kankrin and Tol', the former conveniently retired while the latter died in the same year. Their successor, Count Kleinmikhel – from one of the German families that came to Russia in the time of Peter the Great – was a railway enthusiast, and with his support construction went ahead, subject, however, to the Tsar's insistence on Russia supplying the capital, engineering skills and all requisite materials. In the event the capital could only be raised with the help of German bankers, the engineers – recruited by drafting almost the whole graduating class of 1843 of the Imperial School of Engineering – were only up to constructing the permanent way, and less than 1 per cent of the rails came from the ironmasters in the Urals. The Alexandrovsk State Factory, set up near St Petersburg to build the rolling stock, had to recruit its two managers and many of its craftsmen in the US, while more than 99 per cent of the rails were imported from England.

Local contractors were, however, able to construct the permanent way, working with a labour force comprising up to 50,000 serfs. With the new railway police ready to capture absconders and suppress any unrest, working conditions were appalling, particularly for those assigned to building earthworks in the marshes along the line of rail. So many died that even the imperial bureaucracy was concerned, but Kleinmikhel, anxious to please the Tsar by completing the work on time, ensured that next to nothing was done to improve conditions. He was, however, unable to overcome delays caused by shortage of funds as a result of the Tsar's military expenditure, so that it was only in 1851 that the line was completed. It ran almost dead straight over the distance of some 410 miles separating St Petersburg and Moscow, a result achieved at the cost of substantial bridges and earthworks: the steepest gradient – over the Valdai Hills – was only 1 in 125. Although the civil engineering was of high quality, the trains were appallingly slow, with even the fastest trains – with first-class carriages – averaging only 20 m.p.h. This was partly the result of regular stops in the middle of nowhere, with substantial stations built with buffets and locomotive shops. The sale of tickets was chaotic, giving rise to scenes at the two city terminals such as today are all too familiar to air passengers.[4] The rails could not tolerate high-speed traffic, and the relatively lightweight locomotives – which were all that they could support – in any case lacked the necessary power.

Judged by the volume of traffic, both passenger and freight, the Nikolaev Railway – so named after the death of Tsar Nicholas I in 1855 – was an immediate success. For several years, however, it operated at a loss: one reason was a considerable imbalance of traffic, particularly in agricultural produce – mostly grain and flour – which was almost entirely destined for St Petersburg, either for local consumption or for export by sea; another reason was that rates, for both passengers and freight, were set at absurdly low levels in the national interest. At the same time, Winan, one of the

American managers at Alexandrovsk, exploited his contract by gross overcharging for the maintenance of rolling stock; once he had been bought out, at great expense to the railway, these costs declined from 30 to 15 per cent of gross income. This was but an early example of the sort of unscrupulous large-scale exploitation by private individuals that consistently plagued Russian railway companies – foreshadowing, indeed, similar massive scams in the present-day Russian economy.

In the summer of 1855, war in the Crimea, fought against Britain and France, taught a hard lesson to imperial Russia about the strategic usefulness of railways. For the French and British a successful conclusion to the siege of the main Russian stronghold at Sebastopol, which they had opened in September 1854, was the key to winning the war. In the early months of 1855 the British built a railway from their base at Balaclava on the Black Sea to the siege lines overlooking Sebastopol. With the spring weather the railway was able to bring more than 500 guns and plentiful ammunition to the siege lines, allowing the French and English, on 8 April, to resume the bombardment, which had been suspended during the harsh Russian winter. The new guns were decisive: although the Russians defended Sebastopol throughout the summer, it finally fell in September, so that Russia lost the war.

The government, no longer in any doubt about the need for a major railway construction programme, lost its inhibitions relating to foreign participation and devised a scheme for the massive expansion of the railway network. This depended on setting up the Main Company of Russian Railways to construct, within a period of ten years, four new long-distance lines: one extending the existing network – consisting mainly of the Nikolaev – to the Baltic Sea, a second to a link at Warsaw with Austrian railways, a third to the Crimean Sea, and a fourth east to Nizhnii Novgorod to open up the supply of iron from the Urals. The concession was granted to a French consortium headed by the brothers Pereire of the Credit Mobilier, a bank that already played a key part in financing French railways – which would later include the Paris Metro. The Pereires, managing the company from Paris, milked it of its funds, so that within the agreed time limit only the lines to Warsaw and Nizhnii Novgorod had been completed. Although in 1863 the usefulness of the Warsaw line for sending troops to crush a local rebellion underlined the strategic importance of railways, its construction by the Main Company was still part of a colossal scam by the two brothers in Paris.

The great expansion of Russian railways took place in the fifty-odd years between 1866 and the First World War (1914–18). During this period the total length of track increased fourteenfold, from 3,000 to nearly 44,000 miles.[5] While the greater part of the extended network was constructed during two boom periods, 1868–77 (12,000 miles) and 1893–7 (8,266 miles) the greatest achievement of all, the Trans-Siberian railway, was only completed in the twentieth century.

The Trans-Siberian represented the culmination of a process in which the government, in deciding its railway policy, had to balance strategic and economic needs. The ongoing process of decision, which took several turns in the course of fifty years, had to determine the priorities of different routes, but also who should be responsible for financing and constructing them. Strategically, new routes should support a Russian military presence along every coast and in every frontier area where foreign powers threatened the empire. Since the army could not be everywhere at once, priority both in constructing and operating lines was given to the rapid and large-scale movement of men and material. At

RUSSIAN RAILWAYS IN 1914
N
Samara
Ekaterinsburg
Omsk
Krasnoyarsk
Blagoveshohensk
Khabarovsk
Irkutsk
C.E.R.
Krasnovodsk
Tashkent
Vladivostok
Kushka
Andizhan
0
2000 km
Key
Railway
Present day frontiers
Territory not belonging to Russia in 1914
Archangel
Kotlas
Revel
St Petersburg
Novgorod
Vologda
Vyatka
Perm
Ekaterinburg
Vindava
Libava
Riga
Bologoye
Kostroma
Yaroslavl
Chelyabinsk
Dvinsk
Nizhnii-Novgorod
Kazan
Vilnius
Smolensk
Moscow
Ufa
Minsk
Ryazan
Simbirsk
Tula
Samara
Brest
Gomel
Orel
Penza
Orenburg
Chernigov
Kursk
Uralsk
Zhitomir
Voronezh
Saratov
Kiev
Kharkov
Ekaterinoslav
Tsaritsyn
Kishinev
Nikolaev
Rostov
Odessa
Astrakhan
Novorossilsk
Tuapse
Kislovodsk
Vladikavkaz
Poti
Batum
Tiflis
Krasnovodsk
Baku
Erevan
0
500 km
0
311 miles

Map 8

the same time, given Russian obsession with security, as much as possible should be left in Russian hands. Policy then required new lines to be built to points along the frontiers with Germany and Austria, to the Black Sea and into the mountains of the Caucasus – for defence against Turkey and its allies – and to the states neighbouring Afghanistan to counter the threat from the British in India. Finally, as a result of the conflicting interests of the great powers in China, a railway was needed to serve the Russian treaty ports: this, then, was the beginning of the Trans-Siberian Railway, which became all the more important to Russia after the victory of Japan in the war against China in 1895.

Economically speaking, the important new lines were, on one side, those leading to major port cities, such as Riga on the Baltic Sea and – seen as doubly important after the opening of the Suez Canal in 1869 – Odessa on the Black Sea, and on the other those carrying coal from the Donbass, south of Moscow, and iron from the Urals, east of Moscow, both critical for the growth of a comprehensive railway industry. From an economic perspective, foreign finance was more than just an expedient: without it the new railways would not have been built. Unfortunately for Russia, the investors who followed the Pereire brothers, such as the German, von Derviz – who built the line eastwards from Moscow to Ryazan – were equally adept at plundering the companies set up with dividends guaranteed by the state, although, to do justice to von Derviz, he at least built his line to high-quality standards and had it ready on time. Russian entrepreneurs – such as the egregious Kiev plasterer, Polyakov – in spite of being preferred by the government, were no more scrupulous: if anything their favoured position led to corruption at the highest levels, so that when Count Tolstoi – the minister who had given Polyakov the concession for the Kozlov–Voronezh railway – died, his estate was found to include half a million roubles' worth of shares in the company. Polyakov also kept a little nest egg for himself in a German bank.[6]

While all the fat cats became even fatter, labouring conditions on the railways remained abysmal: even after the abolition of serfdom in 1861, the abundance of cheap labour – helped on occasion by employing soldiers and convicts – was taken for granted. The inevitable showdown came in 1905, with delegates from twenty railway companies combining to set up the All-Russia Railway Union. Encouraged by a small number of Bolshevik activists it organized a general strike in Moscow on 8 October, which by 17 October had extended nationwide. It ended on 21 October after Tsar Nicholas II had granted some of the union demands. In December, a new and more extensive strike, involving workers from twenty-nine different lines, led to violence in Moscow and complete loss of control by the government on the Trans-Siberian railway, where striking railwaymen were supported by mutinous soldiers returning home from defeat in the war with Japan. Troops were used to recover the Kazan railway, with 300 railwaymen dying in the process: even after he won his action, the commander, General Orlov, had another seventy shot and a stationmaster who – at a critical moment – failed to provide a train, hanged. In the years up to the Second World War the Union, having banished the Bolshevik revolutionaries, at the same time extending membership to clerical and administrative staff, became the largest labour organization in Russia.

The Trans-Siberian was the culmination of railway construction in Russia, and its greatest achievement. The line from Moscow to Valdivostok on the Pacific Ocean was one and a half times as long as the Canadian Pacific line from the Atlantic port of Halifax to the Pacific port of Vancouver,[7] but it was never going to be as profitable. As far as western

Siberia the line was useful first for opening up the land to new frontier farmers and then for sending their produce to markets in European Russia, but eastern Siberia – which in Soviet times proved to have vast mineral resources – offered next to nothing before the twentieth century.[8] Beyond Lake Baikal, at the very centre of Russia, the justification for the railways was purely strategic. At the same time the construction programme was extremely haphazard, so that, for example, a long section of line east from Lake Baikal was constructed without any rail link to Vladivostok: this meant that not only the materials for laying the permanent way, but also the initial rolling stock, had to be brought in by river – or in the case of a number of locomotives, by sledge across the frozen Lake Baikal. The lake was always the biggest obstacle in the way of completing the railway, so much so that for many years round the turn of the twentieth century this was overcome by ferries from one side to the other. Construction of the rail link, finally completed in 1904, involved laying a line, with thirty-three tunnels, under the cliffs at the southern end of the lake, and crossing the Yablonovoi mountains at a height of 3,280 feet. Siberia, even then, was not short of unskilled labour: this was provided by the prisons for which that part of Russia had long been notorious.

Although, as originally planned, the Trans-Siberian Railway would be situated entirely in Russian territory, ending up in the Far East by following a line close to the Amur river, the Russian government, in the early 1890s, began to have second thoughts. A line of rail across Manchuria, in northern China, would be shorter by several hundred miles, and would also generate more traffic. Following China's defeat in the war with Japan in 1895, the government welcomed Russian investment. The result was the grant of a concession to the Russo-China Bank to construct the Chinese Eastern Railway (CER): the

Russian company would then operate the line for eighty years, although, after thirty-six years the Chinese government would have an option to purchase it. The Russians, however, set so high a price that exercise of the option was certain to be problematic for the Chinese – which was of course just what the Russians wanted.

In the event the CER was soon subject to set-backs that neither side had envisaged. During the Boxer Rebellion in the summer of 1900 Chinese government troops, far from defending the railway, joined the Boxers in plundering it, so that by the end of the summer only a third of it survived intact.[9] Difficult topography had in any case made the line expensive to build,[10] and although it was finally completed in 1904, the Russians lost almost everything when, in the following year, they were defeated in the war against Japan. According to the terms of peace laid down by the treaty of Portsmouth (1905) the Japanese acquired the valuable southern section of the CER, and by converting it to the European and American standard 4 feet 8½ inch gauge, effectively rendered it useless for Russian traffic – which never played any part in Japanese plans for extending their interests in China. In the years before the Second World War, the new Japanese-owned South Manchuria Railway went on to play a critical part in the Japanese exploitation of China.

As for the Russian government, it was left with no alternative but to go back to the original plan and build a new rail link to Vladivostok entirely in Russian territory. Construction of the new Amur Railway, which began in 1908, proved to be extremely expensive, with such challenges as building a 7,545-foot 22-span bridge across the Amur river – frozen during the winter months – at Khabarovsk. The new railway opened provisionally in 1914, but permanently only in 1916 – the last complete year of imperial Russia. A journey across the entire length of the line then took nine days, which meant that with another two days at sea Tōkyō could be reached in eleven days. For passengers from Europe this meant – at the cost of considerable political hazards – a saving of several weeks over travel by sea.[11] The double-track line was only ready in 1918,[12] by which time Lenin's Bolshevik government was in power.

Chinese railways

The nineteenth-century development of railways in Imperial China was a critical stage in the process by which Britain, France, Russia and the US, later to be joined by Germany and Japan – resorting when necessary to the force of arms – took over the control of the country's import-export economy. This involved a number of wars in the middle of the century in each of which the miserably equipped and poorly led Chinese forces were defeated, first by the British in the opium wars, and then by France, which in 1858 supported Britain in the so-called 'Arrow War'. After every defeat humiliating concessions were exacted from the Chinese. As a result of the Treaty of Nanjing of 1842, they lost Hong Kong to the British, who at the same time were granted unrestricted trading rights in Amoy, Foochow, Ningpo and Shanghai. With British Consuls appointed to all these 'treaty ports' the foundation was laid for the indefinite expansion of trade with China on British terms. This was only the first of a series of unequal treaties with Western powers to which China would be subject for nearly a hundred years. In 1858 the Treaty of Tientsin (Zhenjiang) provided for embassies to be opened in Beijing and for ten new treaty ports. These provisions were confirmed by the 1860 Conventions of Peking, which also ceded Kowloon peninsula – opposite Hong Kong – to Britain.

In 1861, as part of an imperial policy to 'learn the superior barbarian techniques to control the barbarian', an imperial Foreign Office, the Tsungli Yamen, was set up to deal with all the newly established foreign embassies. The five bureaux comprised in it, Russian, British, French, American and Coastal Defence, give a good idea of Beijing's perspective on the outside world in 1861. (Japan and Germany counted for nothing at this stage – a position that would not hold for long.) This was but one reform out of many in the direction of modernization, but since their common objective was more to strengthen the existing order than create a modern state, they did little to equip the empire to deal effectively with the new world on its doorstep.

The institutions of this new world were already making an impact inside China. Railways, while playing a key part in the process, faced two problems. First, at every level – the court, the mandarin bureaucracy and the common people – there was the most profound mistrust of any form of locomotion other than that afforded by the wheelbarrow, the horse and the native boat. Railways, by their very nature, did not easily accommodate to the long-established rules of feng-shui, the Chinese art of geomancy: this was particularly true when their routes were chosen by foreign devils. Even if this problem could be overcome – as it was eventually – the Chinese empire, alone, was quite unable to find the capital to finance new construction. The only solution was then to grant concessions to foreign powers, with all the risks that this involved with feng-shui and other traditional Chinese institutions.

Even so, in the interests of the Western imperial powers China was eventually divided into five different regions, so that along the Yangtze valley and south to the Burmese frontier, the railways were built by Britain, in Manchuria and northern China, by Russia, in Shandong, by Germany, in

Fukien, by Japan, and in the provinces bordering Indochina, by France. Mineral concessions and territorial rights were claimed along the new lines – which might otherwise have been unprofitable – so that effectively the railways, as extensions of the rights acquired in the treaty ports, served to partition China among the great powers.

The first line was proposed as early as 1865 when Jardine Matheson & Co, a leading British company, obtained a concession to build a 'road' between Shanghai and Wusong. Actual construction moved very slowly, and it only became clear in 1875 that there were plans to lay a narrow-gauge railway, with 'small locomotives' along the route of the road. Although the Imperial Government protested in 1876, the line continued to be laid but without the use of the company's steam engine. Instead Chinese workmen used hand trolleys to take material to the railhead, with the trolleys each bearing a red flag, with others carried by workmen both before and behind them. Imperial opposition to the steam engine was withdrawn (and reimposed) several times, but the line was finished as far as Kangwan, 5 miles from Shanghai, and officially opened with the first actual train on 30 June 1876.

Some weeks later, during the construction of the remaining 5 miles, a Chinese workman was knocked down by the engine and killed. Riots followed and the line was closed. The Imperial Government, however, purchased the railway, and reopened it in December 1876. Once the Chinese had made the final payment in October 1877, all traffic was stopped, the rails were torn up and together with the engines and rolling stock were shipped to Taiwan, where everything was dumped into the sea – an event commemorated by the Imperial Government erecting a temple to the Queen of Heaven on the site of the Shanghai Station.

Until the mid 1890s there was little progress in railway construction. The reasons were not only cultural but political. In spite of the steady increase in the number of treaty ports, and of the foreign states to which they had been conceded, imperial China did not accept that it had actually yielded any territory. In 1884, however, France, which in the 1870s had effectively taken over Indochina, closed its northern frontier with China. The message was clear. China's longstanding tributary rights were not recognized. In a land with many Chinese residents, and Chinese as the lingua franca of commerce, the Chinese imperial writ no longer ran.

War followed between France and China, and China was decisively defeated. Worse was to follow in 1894, when modern Japan found a pretext to fight its first foreign war, the matter at issue being the balance of power in the Korean peninsula. The fact that Korea, bordering both Russia and China in the north, was also the mainland state closest to Japan, was crucial. Politically, the country was divided into two factions, with on one side 'conservatives' intent on maintaining the country's isolation, according to traditional Chinese models, and on the other, 'liberals' intent on following the Japanese along the path of modernization and commerce with the outside world. In 1876 a powerful mission from Tōkyō, by establishing a Japanese presence in Korea, threatened longstanding Chinese interests, at the same time sharpening the political divisions within the country. Conflict was avoided as a result of the Treaty of Tientsin of 1885, by which China and Japan agreed, in effect, not to become further involved in Korea.

The treaty, according to their own perception, proved to be disadvantageous to the Japanese, particularly after Russia and France, from the beginning of the 1890s, became interested in extending the Chinese railway system into Korea –

which also involved support for conservative anti-Japanese politicians. Japan was then only too ready to find a *casus belli*. In 1894 the chance came after Korea had appealed to China for help in suppressing the revolt of a native religious sect. When China responded by sending troops, Japan did the same, relying on rights granted by the Treaty of Tientsin. The war which then followed was described by one European observer as 'an encounter between such tactics as were employed by Agamemnon at Troy and those that might have been conceived by Moltke'.[13] Inevitably China lost not only the war, but also another ancient tributary, Korea.

China, following its defeat by Japan in 1895, not only lost to Japan all control of Korea and the surrounding seas, but also a substantial part of its own territory. The Japanese advance only stopped when Tōkyō, subject to considerable pressure from the Great Powers, found it prudent to call a halt. The peace terms, dictated by Japan and accepted by China in the treaty of Shimonoseki on 17 April 1895, left Japan in possession of the Liaodong Peninsula and the island of Taiwan, combined with a strong presence in Manchuria and the southern province of Fukien.

This was too much for Russia, France and Germany, and on 23 April their ministers in Tōkyō combined to give friendly advice to the Foreign Ministry that all Japanese territorial claims on the Chinese mainland should be abandoned. Tōkyō could not resist this 'triple intervention', and with ill grace had to be content with Taiwan, and the Pescadores Islands and an additional indemnity from China. As subsequent events were to show the intervention was far from altruistic: it simply gave Russia, France and Germany considerable scope to exploit China, and in this process the construction of new railways played a critical role. The field was wide open, for at the beginning of the Sino-Japanese War in 1894 there were only 225 miles of railway in the whole of China, of which 166 miles had been built with British money and were under British control.[14] As one historian noted as early as 1907, the Western Powers' 'railways policy ... has been a means to an end, an incident in a larger policy, which can only be described as a policy of colonisation'.[15] Both Russia and France sought to link lines within their own empires – Siberia in the case of Russia and Indochina in the case of France[16] – with the Chinese rail network, realizing a 'grand plan for railways ... under Russian and French control from Vladivostok to Canton'.[17] At one stage there was even talk about a north-south line in China, linking up the Russian Trans-Siberian Railway with the French Yunnan Railway, a project that aroused considerable mistrust in Germany and Britain.

Worse still, China's inability to defend itself against Japan led European powers to claim any number of new concessions – Germany in Shandong, Russia at Port Arthur (Dalian) and France at Guangzhou, with Britain, not to be outdone, gaining a 99-year lease of the 'new territories' opposite Hong Kong. (This proved to be a considerable hostage to fortune, when, in 1997, the expiry of the lease meant not only the reversion to China of the new territories, but of the whole colony.) The Americans, with their own 'factory' – the name given to the local base for all commercial activity in China – at Canton, had also been actively trading with China throughout the nineteenth century.

London and Washington had, however, essentially the same long-term strategy: this was to preserve the territorial integrity of China, with the country open, at the same time, to trade governed by open competition between the countries involved. This was pure self-interest: Britain and America had written the rules of a game that they expected to win, but for the time being other great powers, notably France, Germany, Russia and Japan, were calling the shots.

From a Chinese perspective prospects were abysmal in face of the 'tiger-like voracity' of the foreign powers.[18] Future policy was likely to be determined by the fact that the Dowager Empress Ci-Xi, who had been regent to the Emperor Guang-Xu until he came of age in 1889, helped by court intrigue, once again took over the reins of government. Her pretext was that the young emperor was ill – an ominous indication that he might soon be liquidated. (A strong warning by the British Minister, Sir Claude MacDonald, probably saved his life.[19]) The fact that reform – along the lines that had been so successful in Japan – was the policy favoured by the emperor, makes clear that Ci-Xi was opposed to it. Instead her policy was to rid China of the 'foreign devils' and restore its traditional integrity and isolation.

Finally, the construction of the Tientsin–Yangtze Railway, a project dominated over a long period by the conflicting interests of Britain and Germany, shows how, in the end, China successfully turned back the tide of European railway imperialism, after first having to make any number of territorial concessions – mainly to Germany. True to the form of European imperialism in China, the story begins with one particular incident – in this case the murder of German missionaries in 1897 – being taken as a pretext for a claim to territorial rights to be confirmed by a treaty imposed upon the Chinese. With the Sino-German Convention of 6 March 1898 Germany acquired a 99-year lease of Kiaochow Bay on the Shandong peninsula where it proposed to develop the port of Qingdao as a naval base. In Beijing the German minister confided to his British colleague that 'commercially Shantung was intended to be a German province' – an intention acceptable to neither the Chinese nor the British. To make it come true a company was formed – the Schantung-Eisenbahn-Gesellschaft (SEG) – to build, first, a railway linking Qingdao to the provincial capital, Jinan, and then two more lines, linking these cities to Tongshan in the south of the province. The fact that both Jinan and Tongshan were on the natural route for any rail link between Tianjin (the port for the Chinese capital, Beijing) and the Yangtze river was not lost on the Germans: indeed, such a link would be 'the key to the economic and strategic success of the German railway triangle in Shandong'.[20]

At this stage the Germans hit a snag: in 1897, a Chinese merchant, Yung-win, had already been granted the right to build a railway linking Tianjin and the Yangtze with foreign capital. The German minister, threatening 'serious consequences', obtained from the Tsungli Yamen an 'undertaking that no railway would be built through Shandong without prior German consent; that the projected Tianjin-Yangtze line would run through that province; and finally that German engineers, matériel and equipment would be employed in the construction of this line'.[21] This was getting considerable mileage out of the murder of a few missionaries. Although all this was most unwelcome to the British, both economic and diplomatic factors inhibited them from putting forward any alternative. The best that could be hoped for was a joint venture, based on a consortium formed by the British Hong Kong and Shanghai Banking Corporation (HSBC) and the German Deutsch-Asiatische Bank (DAB) – the source of the SEG's finance – in 1895. When the two ministers in the Beijing legations fell out over this arrangement, the matter went on to higher level diplomacy between London and Berlin.

At the heart of London's policy was the principle that the Yangtze valley belonged to the British sphere of interest. To quote the Prime Minister, Lord Salisbury, the problem was the absence of any 'syndicate really willing and able to build a railway in the Yangtze region'.[22] In short, Britain was no more than a paper tiger. While Germany, on the other hand, had in the SEG an actual company ready to build a railway in China, the Tientsin–Yangtze line was far beyond its resources. Both sides were beholden to the so-called 'myth of the China market'. Germany tried to solve the financial problem by approaching the Russo-Chinese Bank, which was the principal instrument of Russia's *pénétration pacifique*. This made sense in so far as the bank was already involved in the construction of a line north from Tientsin, linking up with the Russian railway network – then being extended to the Far East by the Trans-Siberian Railway. St Petersburg, however, would have nothing of it, so the Germans, unable to go it alone, had to work with the British, who by this time had a syndicate based on the British and Chinese Corporation (BCC), a company associated with the HSBC . By mid July 1898 both the BCC and the German DAB had separately applied to the Tsungli Yamen for the Tientsin–Yangtze railway concession. The Germans were turned down on the grounds that they had already been conceded too much in Shandong province; the British Beijing Legation then informed the Tsungli Yamen that any concession granted would extend not only to the Germans, but also to the Americans – who had consistently refused to recognize any exclusive German right to Shandong. This opened the way to a preliminary agreement, based on a joint British-German project, on 18 May 1899, when the Tsungli Yamen reluctantly accepted a joint proposal in what proved to be 'the last concession granted by China to foreign governments in the "battle of concessions"'.[23]

Both the British and the Germans were far from being out of the wood: progress was all the more difficult because of two signal events in the nine years leading up to 1908, when actual construction started. The first of these was the Boxer uprising of 1900, the second, the Russo-Japanese war of 1904–05. The Boxers – so named because of their martial exercises – constituted a popular protest movement that had originated in Shandong province, partly as a result of the German seizure of Kiaochow. Although the corrupt imperial bureaucracy was its prime target, Christians, both missionaries and their converts, were frequent victims of Boxer violence. When the movement spread to Chihli province, where Beijing was located, the Dowager Empress, almost as a measure of self-defence, encouraged the Boxers to confront the foreign devils: in the summer of 1900 they responded by laying siege to Beijing's foreign legations, an action supported by imperial troops. The expatriate community, led by the ministers plenipotentiary of the powers represented in China – who in a desperate situation were ready to forget that they had continually been at loggerheads over such matters as rival claims to railway concessions – organized the defence of the legation quarter of Beijing. The Boxers, cutting telegraph wires and pulling up railway tracks, soon cut off all communication with the outside world, leaving the legations, with such forces and arms as they could muster, to confront the Boxers alone. The world already knew of their precarious position, and the great powers, with their naval squadrons off the coast of China, lost little time in organizing a relief expedition. The first expedition – led by the British vice admiral, Sir Edward Seymour, but including troops from Russia, France, Japan, Italy, Austria and the US – having commandeered trains belonging to the imperial railway in Tientsin, succeeded in reaching Langfang, about halfway to Beijing and only 30 miles from the city, before they encountered any Boxers. The action, when it came, was ferocious, with imperial troops supporting the Boxers. Further progress was impossible, and after a mainly German force had held Langfang station during five days of almost continuous fighting – in which thousands of Chinese were killed – Seymour had no choice but to retreat to Tientsin. With returning by train impossible after the Boxers had torn up the rails, Seymour's troops had to walk most of the way back to Tientsin, although for the last 20-odd miles junks on the Peiho river were commandeered for the wounded.

Where Seymour had failed in June, a larger and better equipped force succeeded in reaching Beijing on 12 August, to relieve the defenders in the legation quarter. Resistance soon collapsed, the court fled Beijing, and the representatives of the great powers were left to survey a devastated countryside. They left their troops both to guard the important lines of communication and to garrison the legation quarter in Beijing.

The Peace Treaty of 7 September 1901 ensured that the Imperial Government paid a high price for its support for the Boxers. China was to pay a colossal indemnity of £67,500,000 – to be shared among the powers represented at Beijing – spread over forty years. (The last instalment was finally paid in 1940, by which time China was deeply involved in war with Japan.) The legation quarter was to be under foreign control, with no Chinese permitted to reside there, while foreign troops would guard twelve points along the lines of communication between Beijing and the coast. Membership of any anti-foreign society was subject to the death penalty, and the Tsungli Yamen was converted into a true ministry of foreign affairs, the Waiwu Pu. The Dowager Empress was allowed to return to Beijing, where – bowing to reality – she accepted all these reforms.

The directors of the German DAB expected that they would have a much freer hand when it came to building their railway. For one thing, it was the murder of their minister in Beijing, Baron von Ketteler, by a Chinese officer on 18 June 1900, that opened the siege of the legation quarter; what is more, the Peace Treaty required the Chinese to erect a monument on the exact site.[24] For another, the Germans, encouraged by the Kiaochow governor, were busy surveying projected railway lines in North China.[25]

The DAB directors were however to be disappointed: they were turned down by the Waiwu Pu, which rightly insisted that under the 1899 agreement no concessions could be granted without British participation. They had failed to reckon on two key factors: the first was that the reforms following the defeat of the Boxers had encouraged a new spirit of economic nationalism, with Chinese entrepreneurs only too ready to claim their share of the action, and the second, that the British and American ministers supported the Chinese. As the British Minister, Sir Ernest Satow, said, 'we have heard a good deal of la conquête paisable de la Chine par le chemin de fer and that is what I am trying to oppose ... it is necessary for us to be vigilant on behalf of China'[26] – not a sentiment shared by his German counterpart.

The second signal event to affect the negotiations was Japan's victory over China in the Russo-Japanese War of 1904–05. This was critically important for railway politics since it meant that Russia, which had been building railways in Manchuria at a rate comparable to that of the French in Yunnan and the Germans in Shandong, no longer counted. That their place might be taken over by the Japanese was of relatively little interest to Imperial China in its final days: the humiliation of one of the Western great powers counted for much more. Finally, in 1906, the British CCR (Chinese Central Railways) joined up with the German DAB – which by this stage was making concessions to the Chinese, unthinkable in 1899 – and a final agreement for constructing the Tianjin–Yangtze railway was signed on 13 January 1908. This was the signal for construction to start at both ends of the line, with the CCR working north from Pukow on the Yangtze and the DAB south from Tientsin: the two lines linked up in 1912 at the point where the railway crossed the Grand Canal of China. The German SEG had already, in 1910, completed the link between Tientsin and Jinan, but by this time Imperial China was in its final days – largely as the result of a political process in which railways played a critical role.

The Railway Protection Movement, which emerged in the context of the unprecedented freedom of action enjoyed by Chinese political forces after 1905, was born out of the initiative of local commercial and industrial entrepreneurs, who, in their own interests, sought to finance the extension of the railway network, particularly in the rich provinces along the Yangtze river. In purely business terms this would have made sense if they could have raised the necessary capital, but they were far from being able to do so. In 1910 the Minister of Transport, Sheng Xuanxuai, saw foreign loans and outside technical assistance as the only way out of the impasse. For the Railway Protection Movement this meant going back to the bad old days of foreign concessions. In Szechwan the entrepreneurs, who dominated the new provincial assembly, reacted violently against Sheng's proposal. A government boycott was proclaimed and taxes went unpaid. There were mass arrests of demonstrators and local leaders of the movement, and some were killed.

The unrest was not confined to Szechwan, but extended to many other provinces, particularly in the middle reaches of the Yangtze river. Revolutionary movements at provincial level, sharing the common aim of establishing a republic – after bringing the Qing dynasty to an end – often won over local units of the imperial army. In Wuchang, the hub of the planned national rail network, the support of the local military commander was critical for the success of an uprising in October 1911. (It also helped that a part of the Wuchang garrison had been sent to help quash the unrest in Szechwan).

Within two months, the momentum gained as a result of victory at Wuchang led to representatives from seventeen provinces meeting together at Nanjing to form a provisional republican government. Sun Yatsen, who presided over the meeting, succeeded in merging a number of pre-existing revolutionary groups – including the Railway Protection Movement – into a single party, the Guomindang. This then became the government of the new republic proclaimed in 1911. The troubles ahead of it were to come not from Germany – or any other Western power – but from Japan. In 1914, Japan, out of pure opportunism, sided with the Allied Powers in the First World War. By the end of the year it had overcome all the German forces in mainland China, so that effectively the Tsingtao peninsula, and much of Shandong province, became subject to Japanese rule. Since Manchuria, following the defeat of Russia in 1905, shared a comparable fate, during the war years Japan was the power to be reckoned with by the new Chinese republic. Of the great Western powers only Britain, France and the US still counted, but they had much more serious problems on the other side of the world. In November 1914, the Japanese cabinet formally adopted a list of twenty-one demands – the last seven formulated by the General Staff – to be made on China. Their publication in January 1915 led to popular outcry in the US, and under pressure from President Woodrow Wilson the seven required by the Japanese military were withdrawn. None the less, on 25 May – still remembered in China as the 'Day of National Humiliation' – the republic, confronted by a

Japanese ultimatum, signed a treaty accepting the fourteen remaining demands. This meant, to a very large extent, that the history of railways in China after 1915 would be written by Japan. In the years between the two world wars, blowing up trains was all too often an instrument of Japanese imperialism in China. On 4 June 1928, Zhang Zuolin, a troublesome Chinese warlord, was assassinated as the result of a plot by staff officers of Japan's Kwantung army to blow up his train as it approached the Manchurian capital of Mukden. One thing led to another and on 18 September 1931 two colonels of the Japanese Kwantung army general staff, without any authorization from higher levels, orchestrated an explosion at a point on the South Manchurian Railway. As intended, this led first to clashes between units of the Kwantung army and troops loyal to Zhang Xueliang – who had succeeded as the local warlord and whom the Japanese found just as troublesome as his father. Within a year this so-called 'Manchurian incident' had provided the Japanese with a pretext for setting up an independent state of Manchukuo. Its emperor, Pu-yi, two years old when the Dowager Empress Ci-Xi died in 1908, had then succeeded to the throne of an empire that ceased to exist three years later. He was to occupy the throne of the new empire of Manchukuo until the defeat of Japan at the end of the Pacific war in 1945. His final years, spent as a subject of Mao Tse-tung's People's Republic, are portrayed in Bernardo Bertolucci's film *The Last Emperor*.

Railways in Japan

Although in the first half of the nineteenth century, the US, in the Far East, was – together with European powers such as France and the UK – mainly interested in opening up China for commerce, Washington never lost sight of the fact that

Japan was still closed to American citizens. During the 1840s the prospects for maritime commerce changed radically following the introduction of steamships built of steel – as related in Chapter 12. But this raised the new problem of establishing coaling stations, for which purpose, Yokohama in Japan had an exceptionally good location. This – together with continued pressure from the owners of whalers – gave a new impetus to American policy. In 1852 the US Navy Secretary, W.A. Graham, appointed Commodore Matthew Perry to represent American interests at the court of the Japanese *shōgun*. The choice was well made, for Perry was as gifted in diplomacy as he was in organization.

Perry's remit was extremely liberal. Although American interests came first, Perry's instructions made clear that any benefit resulting from his dealings with Japan would be 'ultimately shared by the civilized world'. At the same time he learnt that it was 'manifest from past experience, that arguments or persuasion addressed to this people, unless they be seconded by some imposing manifestation of power, will be utterly unavailing'. Furthermore, Perry was to 'bear in mind that as the President has no power to declare war, his mission is necessarily of a pacific character', and that he was to 'do every thing to impress [the Japanese] with a just sense of power and greatness of this country, and to satisfy them that its past forbearance has been the result not of timidity, but of a desire to be on friendly terms with them'.[27]

Perry spent the greater part of 1853 and 1854 in the Far East, and in both years he visited Japan with his *kurofune*, or 'black ships', powered by steam. The first visit – of only eight days – was largely exploratory, but Perry's diplomacy paved the way for a second visit a few months later, which would be the occasion for negotiating the Treaty of Kanagawa. The United States gained almost everything it had been asking for in Japan, and the treaty, by inducing Japan to abandon '*sakoku*', its self-imposed policy of isolation from the rest of the world – which had been in force for well over two centuries – is rightly regarded as a landmark in the country's history.

Conforming to oriental custom, Perry, on his second visit, brought a variety of gifts, among which was a quarter-size model railway, complete with locomotive, tender and a carriage, with several miles of rails. The American visitors having laid a circular track – about a mile long – behind the reception hall at Yokohama, proceeded to show the assembled dignitaries what the train could do. They were overwhelmed. According to Perry's official record: 'Crowds of Japanese gathered around, and looked on the repeated circlings of the train with unabated pleasure and surprise, unable to repress a shout of delight at each blast of the steam whistle.'[28] One official actually rode the whole circuit, sitting on the roof of the diminutive carriage, and reported that the experience was 'most enjoyable'. Travelling at 20 m.p.h. was far beyond anything conceived possible in what was still a feudal state. As one Japanese railway official noted nearly fifty years later: 'When the railway system was, for the first time, introduced into this country ... we virtually stepped into trains out of [palanquins].'[29]

The arrival of Commodore Perry was no isolated event: his success in opening up Japan to overseas commerce led almost immediately to the major European powers following the US in establishing consular offices with sufficient muscle to ensure the security of their own nationals; these in turn wanted to lose little time in exploiting the commercial potential of Japan. All this took place in a country ruled by a line of hereditary commanders-in-chief, known as '*shōguns*', from their castle in Edo: the Tokugawa dynasty, named after its founder in 1600, maintained a traditional feudal society, whose only contact with the Western world was via a small Dutch trading post on a diminutive offshore island in the bay of Nagasaki. This contact was sufficient to admit Western technology – known as *rangaku*, or 'Dutch science' – on a very limited scale, which, none the less, did have some impact in fields such as armaments. Then in a single year, Commodore Perry, with his steamships and model railway, eclipsed everything that the Dutch had achieved in two and a half centuries. Plainly the regime of the *shōguns* was going to have to accommodate the new technology, and, in doing so, accept what is now called 'globalization'.

This was a tall order to say the least, and in the furthest reaches of Japan local leaders, known as '*tozama*', began planning how to transform the government of the country. Their strategy was to redefine the status of Japan's hereditary leader – known as the 'tenno' according to the home-grown Shintō religion, whose line of descent went back to the year 660 BC, when the Sun Goddess, Amaterasu Omikami, designated Jimmu as the first to hold this office, whatever it might add up to. The issue came to a head when the Western powers forced the opening of the port of Osaka, while the *shōgun* planned to entrust its foreign trade to a guild of twenty merchants. At the critical moment, Kōmei, the reigning tenno, died, to be succeeded by a fourteen-year-old boy, known to history as the Emperor Meiji (1868–1912). In the stand-off between the *shōgun*, Yoshinobu, and the *tozama*, the allegiance of the boy tenno played a critical symbolic role, particularly since Kyōto, where he had his palace, was close to Osaka and far from Edo. While foreign fleets assembled in Osaka Bay for the ceremonial opening of the harbour on 1 January 1868, the *tozama* forces were already advancing on Kyōto, and on 4 January they occupied Nijo, the *shōguns*' palace. Although – given that the *shōguns* hardly ever went to Kyōto – this was a largely symbolic act, it signalled, none the less, the end to a regime that had ruled Japan for 268 years; at the same time little doubt was left in Meiji's court about where its future lay. Yoshinobu left Osaka with its great castle in flames, and after being defeated by the *tozama* at Fushimi on 27 January retreated to his castle in Edo. This in turn surrendered to the *tozama* on 3 May, with Yoshinobu withdrawing into private life.[30] To strengthen their hand, the *tozama* established the tenno and his court in the castle in Edo, and changed the name of the city to Tōkyōto – or 'eastern capital', long known simply as Tōkyō. To the Western world the tenno simply became known as 'the Emperor of Japan', but in fact he was little more than a front for the policies of the Council of State set up by the *tozama*, and the governments that succeeded it. After all, as noted above, the new Meiji Emperor was only fourteen years old when the change took place. The Council of State took over the *shōguns*' administration, and late in 1869 it decided to introduce railways into Japan.

This was a bold decision, and although Japan could never have developed as an industrial power without railways, it was far from being an ideal land for building them. Starting with topography, Japan is an offshore state, consisting of four main islands – Honshū, Kyūshū, Hokkaido and Shikoku – and countless smaller islands. On almost all the islands the terrain is mountainous, and the coastlines indented with innumerable bays and inlets. The largest and most important cities are on the coast, or – like Kyōto – close to it. This has meant, since the earliest days, that travel by water, mostly by sea, was part of everyday life for much of the population. There was no alternative when it came to journeys between the islands, but even so the sea often provided a better thoroughfare for travelling from one part of a large island to another; what is more, Japan, having seen the steamships of the foreign powers anchored at Yokohama, with Commodore Perry's *kurofune* in the vanguard, lost little time in converting inter-island, coastal and river shipping to steam.[31]

Japan is dominated by its largest island, Honshū, in much the same way as Britain is dominated by England. Honshū has most of the key towns, including both the old and new capitals, Kyōto and Tōkyō, and the major industrial city of Osaka, with the port of Kobe just beyond it. Japanese conceive of Honshū – somewhat inaccurately – as having an east–west orientation, with the eastern part, where Tōkyō is, known as Kantō, and the western part, with Kyōto, Osaka and Kobe, known as Kanzai.[32] From ancient times they were linked by a highway, from Tōkyō to Kyōto, known as the 'Tōkaidō', which has a resonance in popular culture comparable to that of the British Great North Road. Almost all human traffic along the Tōkaidō was on foot, with only the occasional officer or feudal lord riding horseback or carried by retainers in a palanquin. Goods were carried by pack animals or human porters. Over a total length of nearly 300 miles there were numerous staging posts, with hostelries for travellers, toll-gates and ferries, so that the Tōkaidō supported its own complex economy, providing a livelihood for thousands; the cultural dimension was equally salient, with shops selling wood-block prints of scenes along the way and books of stories comparable to Chaucer's *Canterbury Tales*. If a railway was to link the towns along the Tōkaidō then much that was central to the character of Japan was certain to be lost. None the less, if there was any traffic artery in Japan that would benefit from conversion to steam, it was the Tōkaidō – whatever the cost to all the innkeepers, toll-collectors, ferry-men and booksellers. Not surprisingly, given all the negative factors, the construction of the railways was a slow process – the more so since Japan had no iron-based industry, nor any institution capable of raising the necessary finance. Inevitably capital would have to come from abroad, and locomotives, rolling stock and rails imported, accompanied by a support team of expatriate engineers and surveyors.

The first line was short, but for all that extremely important. It linked Shinbashi, in Tōkyō, with Yokohama, so joining the capital city with the fishing village that was rapidly becoming a major international port. After successful trial runs in the summer of 1871, the line was officially opened by the tenno in October 1872: he himself rode in a train that covered the 18 miles between Tōkyō and Yokohama in 54 minutes.[33] When the train returned to Shinbashi, large crowds shouting 'banzai' greeted it. Many of the spectators, far from being frightened by the locomotive, were concerned for its well-being.[34] Although the railway soon became a popular symbol of progress and enlightenment, locally, both superstition and economic interests could be problematic. Inevitably the dangers inherent in railway operation soon became apparent: in 1873 sparks from a wood-burning locomotive on the Shinbashi–Yokohama line set fire to several houses alongside the track. A year later, Kanzai, in planning its first railway between Osaka with Kobe – a distance of about the same length – decided, for safety reasons, to locate the Osaka terminus at the edge of the city. Later, in the 1880s, many cities along the Tōkaidō, reacting to the threat to their traditional economy, resisted the new railway planned to link Kantō with Kanzai: this soon proved to be a mistake, for the new employment opportunities created by the railway more than compensated for the loss of traditional occupations. Tradition also explained opposition from samurai, and others tied to the old feudal class system.[35] There was also opposition from members of the new officer class who thought that money was better spent on acquiring modern armaments, but here, as in Europe, the strategic advantages of rail transport for military operations soon won over the opposition. In the wars related above, fought at the end of the nineteenth century on the mainland of East Asia, railways were a key factor.

Somewhat exceptionally – from a worldwide perspective – demand for passenger transport was the driving force behind Japanese railway construction. The market developed very slowly. Given the difficult terrain railways were expensive to build, leading the companies to charge fares beyond the means of ordinary people – so that even on short lines, such as Shinbashi–Yokohama, the early passengers were mostly rich.[36] The breakthrough came in the 1890s after the Tōkaidō line opened in 1889. Then a more realistic fare structure led to a 500 per cent increase in passenger volume.[37] Much of this followed from the long-standing Japanese enjoyment of pilgrimages: not only were trains much faster, but they bypassed the constant shake-down by local officials at checkpoints along the route. By European standards the trains, running on narrow-gauge lines with poor-quality rails, were extremely slow. (At the end of the century a visiting official from India noted how far Japanese railways lagged behind as a result of undercapitalization and obsolescence.[38]) Hard-arse conditions were the lot of lower-fare passengers, and the absence of toilet facilities led to fines of 10 yen for peeing, and of 5 yen for farting, out of the window of a moving train.[39] From 1897, carriages colour-coded to indicate their class left passengers on the Kanzai Railway in no doubt about where they stood.[40]

To begin with the construction and operation of railways, as earlier in Europe and North America, were left to private enterprise. The 1880s – like the 1840s in Britain – was a decade of investment fever, during which many new lines, most notably the Tōkaidō, were opened. There were, however, already cries for state involvement: in the interior of Japan, with its mountainous terrain and low population densities, railways were inherently unremunerative. None the less the defence ministries, which by 1890 had become extremely powerful, wanted such railways for strategic reasons, while local representatives in the Japanese Diet wanted them for the opportunities they provided for graft.[41] 'Pork-barrelling', as much as in the US where the term originates, was – and still is – alive and well in Japan.

In the event it was sericulture that justified Japanese state investment in railways. Historically the combined cultivation of mulberry bushes for silkworms to feed on and the actual production of silk on reels was a cottage industry, located in those parts of Japan where geophysical conditions were most favourable to planting mulberry bushes. This defined a number of remote regions, mostly inland, of which the two most important, the Kiso valley and the Ina basin, lay along possible routes for the Chūō Railway, planned as a state-run enterprise competing with the Tōkaidō line in linking Kantō and Kanzai.

Whichever region obtained the railway would enjoy an enormous boost to its silk-based economy. The reason was essentially technological: any delay beyond a day or two, in reeling the silk inside it, after the cocoon is ready, means unacceptable deterioration. So long as the fastest transport of freight inland was by pack animal, this meant in effect the vertical integration, on a very small scale, of the silk industry in any village that cultivated mulberry bushes. One result was that reeling – an extremely laborious process given that a single cocoon contains several hundred metres of usable silk – relied on technology at the level of the spinning wheel. A girl born in any village given to sericulture could look forward to a lifetime of reeling silk. Not only was this an occupation to which the term 'soul-destroying' hardly does justice, but it was also a considerable bottleneck in the production process.

By the 1890s, when the Chūō Railway was built, new technology made it possible to reel silk on an industrial scale: any sericulture region, therefore, that could ship its cocoons by train – reaching in hours a destination that earlier would have taken days – was going to be in the money in a very big way. With the elimination of the reeling bottleneck, the land planted in mulberry bushes could be vastly extended. In the planning stage a number of alternative routes were considered for the Chūō Railway. From Suwa, in the heart of Honshū's mountainous backbone, to Nagoya, on the coast, one possible route followed the valley of the Kiso river, while another crossed the Ina basin. Both regions had an economy based on sericulture, and their representatives in the Diet fought hard for their constituents. The level of animosity is reflected in the Kiso 'fight song':

> Slaughter all those Ina swine! Don, don.
> So they're not going to let us lay a railroad along
> the Kiso, are they? Don, don.[42]

Although sericulture was more developed in the Ina basis, the Kiso river valley got the railroad. This was in part because it had the backing of both the military and the imperial household, but at the end of the day it was the railway's own directors who preferred a route which, with no gradients more than 1 in 40, would be much cheaper to build. As a concession to the Ina basin a loop was built to touch its northern end, which also had the advantage of avoiding the necessity for a long tunnel at the Sasago Pass. As predicted Kiso boomed, while Ina sericulture went into long-term decline.

The overall result was to create a silk industry that became Japan's leading export earner – with the US as the largest client. This was one reason why the military supported the state-run Chūō line: at a time of imperial expansion on the East Asian mainland, the foreign exchange earned from silk would buy armaments. (Even so, in 1941, the Japanese military accepted that the attack on Pearl Harbor, which actually took place on 7 December, would close, with one blow, its leading export market: in the post-Second World War world it never recovered.)

For Japan's railways, silk, with its inherently high price/weight ratio, was never going to rival coal, which generated, inevitably, a much greater volume of traffic. This is the main reason why silk, from the earliest days, depended on state railways such as the Chūō line. Private enterprise preferred to invest in lines that carried coal. The traffic was essentially short haul, since all that was needed was to bring the coal to the nearest sea-port. (This, after all, was the whole rationale of the Stockton & Darlington Railway at the beginning of the nineteenth century.) The Chikūhō coalfields in the north-east of Kyūshū generated a considerable network of private lines for carrying coal to the seaports of Wakamatsu and Moji: the result was a sixfold increase in production, and a 50 per cent increase in market share. There was no alternative to onward shipment by sea, since Honshū, the main market, was a separate island. (Since 1942 the Kanmon rail tunnel has linked Kyūshū and Honshū, and the line is now part of the Shinkansen system). With the new short-haul railway lines coal became Japan's most important export – after silk and tea – with some 40 per cent of total production going overseas.[43]

Railway construction was slow to lead to Japanese manufacture of locomotives and rails. Until the mid 1880s Japan relied almost entirely on British imports, but by the end of the nineteenth century these had lost out to competition from Belgium and Germany, and above all the US. Only in 1893 was the first home-built locomotive produced by the state railways' Kobe workshop, where the works supervisor, R.F. Trevithick, was the grandson of the Trevithick who in 1894 – as related in Chapter 4 – produced the world's first steam locomotive. The first rails were only produced in 1901, by the state-owned Yamata Iron and Steelworks: within a year, however, it met more than 40 per cent of local demand. State involvement in Japanese railways was the reason behind this success. In the mid 1900s this extended to outright railway nationalization, carried out so as to give Japan a comprehensive national network – known even in the country itself as JNR, for 'Japanese National Railways'.

The benefits of nationalization were mixed. In south-west Honshū, along the coast of the Inland Sea facing the island of Shikoku, the new San'yō line, created out of the merger of several local private railways, achieved very high standards. From 1899 all carriages were fitted with electric light; express trains, running to US standards, had already been introduced in 1895 – in 1899 they acquired dining cars and in 1900 sleeping cars. Better still, an official visit to Britain in 1899 led to the San'yō railway requiring its employees to observe British standards of politeness. The Kanzai railway, which linked up San'yō, to enable through traffic from the Tōkaidō line, was next best: one reason was said to be the need to compete with the comfort and efficiency of the steamer services on the Inland Sea.[44]

The JNR correctly foresaw a great future for suburban railways, which accounted for thousands of kilometres of new track: electrification started in 1912 to extend very rapidly thereafter.[45] Cheap fares attracted such numbers to commuter services that they became notorious for the way passengers were packed into the carriages. Urbanization – on an unprecedented scale – was then the name of the game, but all this was mainly the world as it was after the First World War, and hardly belonged to the age of steam.

12

STEAM CONQUERS THE OCEANS

In April 1808 John Stevens, a New York shipbuilder, having launched the *Phoenix*, a steamboat intended for Philadelphia as a passenger ferry on the Delaware river, decided to bring it to its destination by sea along the coast of New Jersey. With a captain uncertain of its seaworthiness, the *Phoenix*, after several mechanical breakdowns at sea, reached Philadelphia after a voyage of thirteen days.[1] Even though it was never again to venture out to the open sea, it had still made history as the first steamship to complete an ocean-voyage.

A comparable story can be told about the first steamship to cross the English Channel. In 1814 the *Margery*, a side-wheeler, was launched in Glasgow for service on the Clyde, but it was soon sold to a London concern to be used on the Thames. In January 1815, with its paddle wheels removed, it was first towed through the Forth–Clyde Canal, to continue its journey under sail – down the east coast of Britain – to

London, where, with its paddle wheels restored, it introduced steam transport to the capital city. A year later a French company bought it for the Seine, renaming it *Elise*: this time round it proceeded under steam, first to Newhaven, and then, in heavy seas, across the English Channel to the mouth of the Seine and on to Paris, which it reached, after a voyage of twelve days, on 29 March 1815.[2]

The first regular steamboat service at sea opened in 1819 when one of Scotland's steamboat pioneers, Henry Bell, transferred the *Comet* – built to his design by a Glasgow shipyard – from service on the Clyde and Forth to a new sea link between Glasgow and the Western Highlands. After seven years' river service the *Comet* only survived one year at sea, before being wrecked off the coast of Argyll on 13 December 1820.[3] The ship was still aptly named, for the trail it blazed was soon followed. Another Scotsman, David Napier – who had designed the *Comet*'s boiler – built the *Rob Roy* for the first regular steamship service across the Irish Sea between Greenock and Belfast, and after two years transferred it to Dover to start the first cross-channel service.[4] With an average time for crossing from Dover to Calais of about two hours forty-five minutes the overwhelming advantage of steam power for short sea journeys was indisputable, and during the 1820s new shipyards in London built steamboats not only for the Channel service, but for routes along the English coasts. Once again the precedent was established by David Napier, who had opened a regular service linking Liverpool, Greenock and Glasgow in 1822.

Napier was also one of the first to realize that a steamboat behaved quite differently at sea from a vessel under sail. While with a steamboat there was no need to tack, it was essential to limit heeling to prevent the paddle wheels coming out of the water. After working first with model boats in a specially constructed tank, Napier produced a new design of hull, which was used not only for his own boats, but for those built by other British shipyards.[5]

History records – somewhat remarkably – an Atlantic crossing by a steamship in the year 1819: the *Savannah*, however, was essentially a sailing ship with an auxiliary steam engine. This hybrid, as its name suggests, was built for passenger service along the coastline of the southern states of America. Here the engine, which was extremely inefficient, would be used mainly for getting the ship in and out of harbour. The ship was a business failure, so its owners decided to try their luck and sell it in Europe. With its passenger accommodation filled with 75 tons of coal and 25 cords of wood, it left its home port, Savannah, for Europe on 24 May 1819. Although its steam engine was only in use for three and a half days, it still ran out of fuel off the coast of Ireland: when it finally arrived in Liverpool it had been at sea for over twenty-seven days. There were no buyers in Europe, so after calling in on Copenhagen, Stockholm and St Petersburg, the *Savannah* returned to its home port entirely under sail – to save the costs of coal. Sold at auction to buyers who stripped it of its engine, the *Savannah* was lost in a gale off Long Island in 1820.

The ignominious history of the *Savannah* points to the key problem inherent in the use of steam for long-distance ocean voyages. In these early days no ship was able to carry the fuel required by continuous steam power for the whole period of a voyage. Although during the 1820s a number of hybrids, much better designed than the *Savannah*, crossed the Atlantic successfully with the 'adroit use of steam and sail',[6] voyages under continuous steam power were confined to short sea-crossings for which only a small quantity of fuel was needed. For obvious geographical reasons this factor enormously favoured Britain, which was further helped by having abundant coal.

What worked in the seas around Britain was equally suited to any part of the world where distances at sea were relatively short. In 1830 this led the British Admiralty, which after the Napoleonic Wars became responsible for postal services to British possessions in the Mediterranean, to replace sail with steam. The first such steamship, HMS *Meteor*, having left Falmouth on 5 February, was back home from its final destination, Corfu – then, somewhat remarkably, a British possession – after little more than a month: this was a considerable improvement over the three months taken by the fastest sailing packets.[7] In 1835 two Londoners and a Dubliner joined forces to set up, under the name Peninsular Steam Navigation Company, a regular steamer service between London and the Iberian peninsula. On 22 August 1837 the first commercial contract for carrying mail by sea was agreed with the Admiralty, and on 31 December 1840 this was followed by the incorporation by Royal Charter of the Peninsular and Oriental Steam Navigation Company (P&O).

Also in 1837 – by which time the Admiralty had thirty-seven steamboats – it set up a special steam department, although steam was not popular with officers of the Royal Navy, whose votes in the Royal Yacht Squadron had been sufficient, in 1829, to pass a resolution excluding from membership any gentleman owning a steam-powered vessel. Faced by this resolution, a certain Assherton Smith resigned his membership and went on to own no less than nine steam yachts. In this he was alone until, in 1856, the Squadron relaxed its rule.[8]

In the course of the 1830s the regime established for the Mediterranean was extended to the Far East, with great advantage to the British in India and the Dutch in the East

Indies. The big problem was supplying fuel, so that in 1837, for example, the *Bernice* spent twenty-five days in harbour coaling out of a voyage of eighty-eight days from Falmouth to Calcutta. Although this was welcome neither to the Admiralty nor to ships' crews – who found the process dirty and laborious – it was none the less a price that had to be paid.

The problem, although formidable for ships bound for destinations – such as those in the eastern hemisphere – where there were sufficient ports of call along the prescribed routes, was daunting when it came to destinations in the western hemisphere. What is more, by the 1830s these were by any standard much the most important – as was well recognized in both Europe and America.

By this time shipbuilders and marine engineers, such as David Napier and his cousin and competitor, Robert Napier, were beginning to realize that for steam to conquer the world's oceans, naval architecture had to be fundamentally rethought. Engine design, material used in construction and means of propulsion all required rethinking. Even so, the Canadian-built *Royal William*, which in 1833 became the first ship to cross the Atlantic – from Halifax, Nova Scotia to Cowes on the Isle of Wight – under steam, showed little radical change in the state of the art. The journey lasted eighteen days which brought the ship very close to the limit of the amount of coal that it could carry; what is more, one day in four had to be devoted to descaling the boilers, an operation made necessary by relying on sea water to provide the steam. At the same time, by travelling eastwards the *Royal William* had the advantage of wind and current: it never made the return journey to Canada, which was almost certainly beyond its power.

The success of the *Royal William*, however qualified, opened the competition for establishing a regular trans-Atlantic steamship service. Robert Napier, whose Govan shipyard had built the crankshafts for the *Royal William*, was ready to take up the challenge, but could not find the necessary finance. However Isambard Kingdom Brunel, chief engineer of the Great Western Railway Company – whose achievements are recounted in Chapter 7 – persuaded his board to set up a separate company, the Great Western Steamship Company, with the specific purpose of transporting passengers across the Atlantic.

Brunel, although no shipbuilder, had the fundamental insight that a substantial increase in size was the key to building a ship that could carry sufficient coal for crossing the Atlantic in either direction. Quite simply, with the increase of the dimensions of a vessel by any factor, its carrying capacity increases by the cube of that factor while the resistance to be overcome by its engine increases only by its square. With the help of this principle it is possible to determine the minimum dimensions of a steamship able to carry sufficient fuel for a voyage of any prescribed length – such as the distance involved in any Atlantic crossing.

Brunel, in applying this principle to the first ship – appropriately named the *Great Western* – to be built by the new company, conceived of a vessel built mainly of wood, on the model of the Royal Navy's ships of the line. Given the need for carrying sufficient fuel for an Atlantic crossing, the ship had to be much larger than any constructed on this model. The need to withstand the force of Atlantic waves also meant that it had to be much stronger. Brunel's design of the hull therefore provided for the standard oak ribs to be joined by four rows of iron bolts, 1.5 inches in diameter and 24 feet long, extending the whole of the ship: following naval practice it was sheathed in copper below the waterline. The *Great Western*, built accordingly and weighing 1,320 tons, was then launched at Bristol on 19 July 1837.[9]

By this time two other shipping companies, one in London and the other in Liverpool, were challenging the *Great Western*. In London the British and American Steam Navigation Company, having commissioned its own ship – to be named the *British Queen* – only to find that the shipyard could not be ready on time, chartered the *Sirius*, a steamship recently constructed for service between London and Cork. Both these ships were equipped with a surface condenser designed by the inventor, Samuel Hall, in 1834, and first installed in the paddle steamer *Prince Llewellyn* which operated a local service out of Liverpool: coupled with a 'steam saver' this enabled ships to steam indefinitely on fresh water, so avoiding the necessity for descaling at sea. This was a key invention in the development of long-distance ocean travel, but even so the continuous circulation of fresh boiler water brought its own problems of corrosion, which were not satisfactorily resolved until the late 1860s.[10]

In Liverpool, the Transatlantic Steamship Company, which was also challenging the *Great Western*, chartered another ship built for the Irish Channel, confusingly named the *Royal William* – given that this was also the name of the Canadian ship that had completed the west-east Atlantic crossing in 1833. It was, however, not ready in time to join the race with the *Sirius* and the *Great Western*.

On 28 March 1838 the *Sirius* left London for Cork, where it would take on fuel sufficient for completing the voyage to New York. This gave it a head start since the *Great Western* was still undergoing trials in the Thames estuary. It finally left Gravesend for Bristol – where the passengers were waiting to go on board – on 31 March, but it was then further delayed by a small fire in the boiler room. The ship's captain, concerned by the amount of smoke, headed for Canvey Island, where he ran it aground. At the same time, Brunel – on board for the maiden voyage – had gone down to the boiler room to investigate, and in doing so fell from a ladder. His injuries meant that he had to be taken ashore, and when the ship was able to proceed with the next high tide, he was not on board – and so missed the maiden voyage across the Atlantic. The *Great Western* finally made it to Bristol on 2 April, by which time most of the passengers, having learnt of the fire, had cancelled their bookings. Delayed first by the need to take on coal and stores, and then by bad weather, it was only at 10.00 a.m. on 8 April that the *Great Western* began its voyage to New York.

Although the *Sirius* had left Cork on 4 April on a notably shorter voyage, it was by no means certain that it would arrive in New York ahead of the *Great Western*. Mid-Atlantic headwinds were so strong that a number of passengers begged the captain to turn back: refusing to do so, in the final hours he could only maintain a full head of steam by burning some of the ship's wooden spars. After more than eighteen days at sea, the *Sirius* finally arrived in New York on 22 April with only 15 tons of coal left. The *Great Western* arrived early the next morning, having taken just over fifteen days – more than three days less than the *Sirius* – to cover a much longer distance at an average speed of 8.8 knots. What is more, there was still more than 200 tons of coal in its bunkers.[11] The *Royal William* was never in the race: none the less it later became the smallest passenger steamship ever to make the Atlantic crossing in both directions.[12]

Vast crowds on the foreshore of New York harbour welcomed both the *Great Western* and the *Sirius*, but the *Great Western*'s welcome when it arrived back in Bristol – carrying sixty-eight passengers, with no cancellations – was even more enthusiastic. Brunel's initiative won for Britain the dominant position on the most profitable passenger shipping route worldwide. The US – ahead in the early days of river steamships described in Chapter 5 – was hardly a competitor when it came to the construction and operation of ocean-going vessels, preferring to give priority to steam on railroads and inland waterways.

Brunel, true to character, could not rest content with the success of the *Great Western*. His next ship, the *Great Britain*, was not only larger, but built of iron rather than wood and driven by a single large screw propeller, which the original inventor, Francis Pettit Smith, had developed for the Admiralty in the late 1830s. Its first successful sea trial – with the specially built *Archimedes* in 1839 – was then followed by the launching, in 1841, of the first warship to incorporate it. Brunel, having noted the potential of Smith's screw propeller, persuaded the Admiralty to measure its potential by adopting it for a new frigate, HMS *Rattler*, whose performance would then be compared with that of a sister ship, HMS *Prometheus* equipped with paddle wheels. The trials, which took place in October 1844, showed that the *Rattler* had an average speed over a measured distance of 9.9 knots whereas that of the *Prometheus* was only 8.75 knots.[13] At this time the *Great Britain*, although launched on 19 July 1843, was still being fitted out in the Bristol City Docks, and Brunel was able to take advantage of the Admiralty tests in adapting Pettit's screw propeller to his new steamship.[14] The engine that would drive it was the largest ever built. The Admiralty for its part was so impressed by the *Rattler*'s performance that it ordered six new propeller-driven iron-built warships.

Although the *Great Britain* was built as a sister ship to the *Great Western*, it was – at 3,018 tons with an overall length of 322 feet – much larger. After the first stage of fitting out in Bristol, in December 1844 – to complete the process – the *Great Britain* sailed to Liverpool which became its home port on the Great Western Steamship Company's North Atlantic service. Two years later, in the late summer of 1846, after running aground on the coast of Northern Ireland, it had to wait until the following spring – exposed to winter storms – before it could be salvaged. Brunel castigated his colleagues for their neglect, but the incident, which was more than the company could afford, led to its liquidation and the *Great Western* was sold.[15] The ship, after several refits, continued to operate for many years in different parts of the world; it only returned to Bristol in 1970 – to become a floating museum – after being rescued from an anchorage in the Falkland Islands where it had been abandoned long before.

Of the two companies that successfully challenged the Atlantic in 1838, this left the British and North American Royal Mail Steam Packet Company – successor to the British and American Steam Navigation Company – as the only one surviving. This company, formed in 1840 by Samuel Cunard – later to become the Cunard Steamship Company – was the first to operate a regular transatlantic service. In 1838, Cunard, who had won from the British Post Office an exclusive contract for carrying mail between Britain and North America, ordered four large ships – the *Britannia*, *Acadia*, *Caledonia* and *Columbia* – from Robert Napier's Clyde shipyard.[16] They were, remarkably all paddle wheelers, and indeed the *Scotia*, the last of this kind to be ordered by Cunard, was only launched in 1861. Cunard, unlike Brunel, was not an innovator in steamship technology: it was only in 1855 that he ordered his first iron vessel, the *Persia*, from Robert Napier. Even so, he was in the long run a more successful businessman and well before the end of the nineteenth century Cunard had become the name that counted on the North Atlantic.

Brunel, for his part, did not give up building ships: in the 1850s – the last decade of his life – a leading Thamesside shipbuilder, John Scott Russell, built the ship ultimately known as the *Great Eastern*, to his design. Its origins are to be found in Brunel's sketch books of the early 1850s, a period in which he was involved in numerous other engineering projects on dry land – including, notably, the Great Exhibition of 1851. He returned to actual shipbuilding in 1852, after he had persuaded the Eastern Steam Navigation Company to commission the *Great Eastern*. This company, only recently set up, was formed to compete with P&O for traffic to India, and the ship proposed by Brunel promised to outperform – in speed, size and comfort – any possible rival.

According to his own account in 1851, Brunel's design was intended 'to make long voyages economically and speedily by steam [which] required the vessel to be large enough to carry the coal for the entire voyage at least outwards and unless the facility for obtaining coal was very great at the out port – then for the return voyage also'. He was convinced that 'vessels much larger than now built could be navigated with great advantages from the mere effects of size'.[17] His vision was realized with the *Great Eastern*, an 'iron ship double-skinned below the water-line and compartmentalized by ten lateral bulkheads running the length of the ship, [with] power derived from two sets of steam engines – one to drive giant paddle-wheels and the other to power a screw propeller'. The ship, in its 'final form [had] five funnels and six masts for auxiliary sail-power [and was] 692 feet long ... with a gross register of 18,915 tons'.[18]

Sadly Brunel's ideas were in advance of their time, though – as subsequent history now shows us – only by a margin of less than a generation. The immediate result was that the *Great Eastern* was built under crippling financial constraints, so that Brunel fell out with both Russell, who was building it, and John Yates, company secretary of the Eastern Steam

Navigation Company. It was little consolation to him that construction of the ship attracted enormous publicity, so that *The Times* saw its launch in November 1857 as justifying the British claim to 'the moral superiority of the world'. It spoke too soon. The launch, which under pressure from the directors of the Eastern Steam Navigation Company took place earlier than Brunel had wanted, failed. It succeeded only on 31 January 1858 after Brunel had seen to the installation of hydraulic presses to push the ship, as planned, sideways into the Thames from Russell's yard at the Isle of Dogs.

Fitting out the ship, once it was in the water, was only complete in September 1859, the month in which Brunel died. He had been ill for a whole year, and two long holidays in Egypt, recommended by his doctors, had failed to cure him. Brunel lived just long enough to inspect the *Great Eastern* two days before it finally set sail on its sea trials: in the late afternoon of the third day at sea a 'terrific explosion hurled the first of her five funnels into the air ... and a stream of boiling water cascaded into the paddle-engine boiler room' leading to the death of five stokers. This, in itself, was an emergency that Brunel had provided for: the propeller boiler room was able to power not only the propellers but also the paddle wheels. The *Great Eastern* continued its westward voyage, drawing into Portland Harbour the next morning, to remain at anchor there while an inquest was held at Weymouth. By the time this was complete, on Monday, 19 September, Brunel was dead. The coroner and jury, after hearing witnesses mainly concerned to blame others for the disaster, were unable to find any single person responsible. The situation on board had been anarchic in a way which Brunel would never have tolerated, and on key matters his specific instructions had been ignored. As presented by *The Times*:

> The immense vessel was a microcosm with internal politics as diverse as those of the United States on the eve of a Presidential election ... we should like to know who the engineers were whose want of caution is vaguely indicated by [the] jury as the cause of a catastrophe that has consigned several of our fellow-creatures to a horrible death, and has cast great gloom over the inauguration of an enterprise which our hopes and sympathies, and pride had made national, and which we had accepted as an exponent of the vigour, enterprise and grandeur of conception, that justify our claims to the moral superiority of the world.[19]

The *Great Eastern*, after all its trials, entered into transatlantic service between Southampton and New York in 1860, but because of its history of misfortune and high operating costs it never achieved commercial success. This only came when it was adapted in 1865 to lay under-sea cables, a purpose for which, although not so designed, it was well suited. Its performance was further improved in 1867 with the installation of a steam steering engine during a periodic overhaul. The inventor, J. MacFarlane Gray, was British, and although others – including the American, Frederick E. Sickles[20] – were earlier in the field, his engine was the first efficient enough to be adopted by shipbuilders, often with its own separate boiler. Powered steering later proved essential for the vastly increased scale of shipbuilding from the 1880s onwards. The same was also true of another invention of the 1860s, the mechanical telegraph system invented by William Chadburn of Liverpool, which provided the rapid and reliable means of transmitting orders from the bridge to the engine room.[21]

The *Great Eastern* represented both Brunel's successes and his failures. Iron ships and screw propulsion became standard (although by the end of the nineteenth century iron would be replaced by steel), but Brunel never found a satisfactory engine: his atmospheric steam engines showed little advance beyond what James Watt had already achieved in the eighteenth century. If he had lived only another year or two, he would have been able to install in his ships the compound engines developed by John Elder and others, and would surely have done so. Elder, who had been an apprentice of Robert Napier, took out his first patent for using such an engine to power a screw propeller in 1853, but Brunel never realized its potential: indeed it probably escaped his notice. But this, as the following paragraphs show, was where the future lay, for 'although it was not realised at the time, this was one of the really important advances in marine engineering, and despite the fact that it did not represent any fundamentally new approach, its effect on the future of shipping and shipbuilding was enormous'.[22]

Although the principle of the compound engine can be traced as far back as a patent granted to Jonathan Hornblower in 1781,[23] practical application was not possible until the development of boilers strong enough to contain superheated steam at high pressure. These two essential attributes of the engine go together: superheating, which means that steam is produced at a temperature higher than that of the boiling-point of water, is only possible under high pressure. Inherent in this process is the absence of any water condensate in the cylinder. The compound design provides for two cylinders, with steam from the first, operating at high pressure, being released into the second, operating at low pressure. The two then combine to produce the required power. Maximum efficiency depends upon achieving the right balance between the dimensions of the two cylinders, in terms first of diameter and length of stroke, and then, derivatively, of pressure and temperature – both of which change in the course of every stroke according to the so-called 'gas laws'.[24] Expanding the steam twice – known as double-expansion – meant a very considerable saving of fuel, at the level of 30 to 40 per cent. Another advantage was the increased regularity of the turning moment, which in turn meant a more efficient propeller.

Although the first British trial of an engine using superheated steam – with the converted paddle steamer, *Dee*, in 1856 – was successful, it was only in the 1860s that such engines became sufficiently reliable for use in ocean-going vessels. Compound engines were first adopted by the Pacific Steam Navigation Company, whose directors wished to save on the cost of coal in South America. For much the same reason, P&O, with its extensive services to India and beyond, soon followed with ten new ships introduced in the five-year period, 1861–6. As for transatlantic services, Cunard, in 1868, had compound engines installed in the *Batavia* and the *Parthia*, by which time the company had decided definitively in favour of propeller propulsion, after the *Scotia*, with its paddle wheels, had lost the Atlantic blue riband to the *City of Paris* in 1867.[25] From 1870 onwards almost all other lines followed Cunard in installing compound engines – so much so that by 1875 they powered more than 2,000 vessels.[26]

The Admiralty had also noted the development of John Elder's compound engine. Given that the Royal Navy's operations in distant waters had come to depend on coaling stations throughout the world – as often as not in places such as Cape Town, Aden, Trincomalee, Singapore and Hong Kong, recently incorporated into the British Empire – that were mainly supplied by coal bought from Britain, any possible economy was welcome. Finally, in 1865, a trial was held by comparing the performance of three identical propeller-driven ships sailing from Plymouth to Madeira: somewhat ignominiously running low on coal forced all three to abandon the race, but even so, the performance of HMS *Constance* – the only one with a compound engine designed by John Elder – made clear the superior efficiency of this type of engine.

Already in the mid nineteenth century, two wars, one in Europe and the other in America, had greatly influenced the development of warships, and derivatively, marine technology. The first was that fought in the Crimea, in the years 1854–1856; the second, the American Civil War of 1861–65. In the Crimea both the French and the British – allied in war against the Russians – learnt the advantages of iron armour. This was first adopted by the French Navy after its wooden ships had suffered heavily from the Russian shore batteries at Sebastopol, but the Royal Navy soon followed on with two iron-armoured floating batteries, the *Erebus* and *Terror*. After the war the French Navy reconstructed the wooden battleship *Gloire*, incorporating wrought-iron plates along its entire length. The Royal Navy went further with the *Warrior* and the *Black Prince*, entirely new line-of-battle ships built with iron hulls and armour in some places 4½ inches thick – rendering 'obsolete virtually every other warship in the world'.[27]

The best-known naval action of the American Civil War was the battle between the southern CSS (Confederate States Ship) *Virginia* and the northern USS (United States Ship) *Monitor*. The *Virginia* is still, however, known to history as the *Merrimac*, the name it had as a ship of the US Navy. This was scuttled and burned when Union forces, in April 1861, had to abandon the Norfolk Navy Yard in the southern state of Virginia. Salvaged by the Confederate Navy, it was rebuilt with a penthouse superstructure covered with two layers of 2-inch-thick iron on the upper deck. Ten 7-inch rifled guns were placed so as to be able to fire through holes in the superstructure, and a heavy iron ram was fastened to the bow. On 8 March 1862, its first day in action, the *Merrimac* rammed a wooden frigate, the USS *Cumberland*, which sank with the loss of many lives in Hampton Roads at the entrance to Chesapeake Bay. It then went on to bombard another wooden frigate, the USS *Congress*, which burst into flames leaving its crew no choice but to surrender. By this time the other Union ships had pulled inshore, where the *Merrimac* – with a draught of 22 feet as a result of its heavy armour – could not follow them.

Union intelligence was in fact long aware of the reconstruction of the *Merrimac*, so that its own ironclad ship, the USS *Monitor*, was built on much the same principle, with two 11-inch guns in a single revolving turret. Congress had first voted funds for researching the possibility of steam-powered warships in 1834 and the first ironclad warship, the USS *Michigan*, a steam-powered side-wheeler, had been launched as early as 1843 – this was the sort of ship that Perry, as recounted in Chapter 11, sailed to Japan in 1853 and 1854, causing consternation to the local inhabitants when it appeared in Tōkyō Bay. In 1861 Congress voted $1,500,000 for propeller-driven ironclads, which then became standard. Early in 1862, however, the *Monitor* – unique in this class – was the only warship ready and able to confront the *Merrimac*. Barely seaworthy after the most summary of sea trials it was sent down to Hampton Roads as soon as the US Navy received the news of the devastation caused by the *Merrimac*. The two ships engaged in battle, at very short range, on 9 March. Although neither side gained a decisive advantage, the *Monitor*'s armour withstood all the *Merrimac*'s shells, while the *Monitor*'s 11-inch guns drove in the *Merrimac*'s armour by several inches leaving the gun crews stunned with concussion. Both sides withdrew at the end of the day, never to meet again. Lessons were learnt from the engagement worldwide. In particular, the *Monitor*'s revolving turret became a standard for iron, and later steel warships, and it was noted also that the propellers driving both the *Monitor* and the *Merrimac* suffered no damage. This was a major factor in the abandonment of paddle wheels, not only for warships but for all other ocean-going steamships – and in the remaining years of the Civil War both sides used propeller-driven ironclads.

The Confederate States adopted a further innovation, which became standard well before the end of the century. This was the twin-screw ship with two propellers. Although first tried out in the 1850s in ships with single-expansion engines, the rate at which coal was burnt was quite unacceptable. In the early 1860s a London shipyard built twenty twin-screw ships with compound engines to be used by the Confederate States as blockade-runners, taking advantage of their high speed and manoeuvrability. Such was their success that the US Navy had to build special new ships to counteract them: the Navy's chief engineer, Benjamin F. Isherwood, also adopted twin-screw propulsion but not the compound engine. His innovative design of the hull meant that his ships were extremely fast – with one, the USS *Wampanoag*, attaining a record speed for a warship of 16.7 knots – but their extravagant fuel consumption led to their being laid up as soon as the war was over.

More generally, technological progress during the 1860s revolutionized the design of ships and the materials out of which they were constructed. The new compound engines, operating at much higher speeds, were only possible because of the development – starting with an American patent granted in 1839 – of a new tin alloy for large marine engine bearings: for smaller bearings another alloy, known as phosphor bronze, was introduced at much the same time. Although the new improvement in quality came with a high price, this was largely offset by economies of scale made possible by the increase in the numbers and tonnage of the new ships. Shipyards, particularly in Britain, had long order books, helped considerably, in the year 1869, by the opening of the Suez Canal. The new short route to India and the Far East was only possible for steamships.

The introduction of steel construction was not without problems, if only because steel-making on a large scale was itself an infant industry. By the late 1870s, with most problems solved, the Admiralty ordered two steel warships, the *Mercury* and *Iris* in 1876, while in 1879 William Denny's Clydeside yard laid down the first steel hull for a merchant ship, *Rotomahana*, ordered from New Zealand.[28]

The year 1876 was also the first one in which more ocean-going steamships were built, worldwide, than sailing ships. For this there were a number of reasons. The leading builders and operators of sailing ships were based in the United States. The geographical position of the Americas in the western hemisphere meant much longer sea voyages, so that the costs of carrying sufficient fuel were high in comparison to those of European shipping. At the same time commercial whaling, which required ships to be months away from harbour, with little advantage to be gained from speed, was mainly carried out by American ships with home ports in New England. Average sailing distances had also increased enormously as the Atlantic – through shortage of stocks – lost out to the Pacific. The Pacific was also critical to the US for the China trade, carried out by the famous tea-clippers. Both the last tea race, and the last sailing of a British clipper, took place in 1872.[29] While the American clippers continued to sail to the end of the nineteenth century, whaling declined very rapidly after the 1870s when electric lighting and the exploitation of oil wells drastically reduced the need for whale oil in lamps. This development was foreshadowed as early as January 1862, when the first oil export shipment from the US to Europe – consisting of 1,329 barrels brought from Philadelphia in the *Elizabeth Watts*, a 224-ton brig – reached its destination at London's Victoria Dock.

The 1870s also witnessed many other key innovations. Britneff, a Russian shipowner, introduced the first steam-powered ice-breaker – named prosaically, in German, as *Eisbrecher I* – in 1871, and in 1872 a Tyneside yard built the first steamship, the *Vaterland*, for the bulk transport of oil.[30] Following the success of MacFarlane Gray's steam-powered steering, from 1873 onwards small steam engines were installed for operating capstans, gun turrets, hoists, bilge pumps and numerous other types of equipment. In 1876 two British warships, HMS *Minotaur* and HMS *Temeraire*, were equipped with steam-driven direct-current generators for electric arc searchlights – a precedent soon adopted in the fitting out of all large warships. In 1879 electric arc lamps were installed in the *City of Berlin* for internal lighting; two years later, in 1881, the incandescent filament lamp – whose invention is credited to two prolific inventors, the Englishman, Joseph Swan, and the American, Thomas Edison – were installed in HMS *Inflexible* and two passenger liners, the Inman Lines *City of Richmond* and the first all-steel Cunarder, *Servia*.[31] The year 1879 was also the one in which the first refrigerated ship with a cargo of meat left Sydney for Britain.[32]

The success of the compound engine from the 1860s onwards was soon followed by that of the even more powerful and efficient triple-expansion engine. Simply stated this involved little beyond adding a third intermediate cylinder to the high- and low-pressure cylinders of the compound engine. The main requirement of the three-stage system was a boiler capable of withstanding even higher pressures than those required by the compound engine. This result was first achieved in 1873 by a triple-expansion engine with water-tube boilers installed in the shipbuilder, Alex Kirk's *Propontis*, but it was only in 1881 that Robert Napier – by persuading the Aberdeen line that steel made by the Siemens process was of a sufficiently high standard for the boiler – was able to incorporate such an engine in the flagship *Aberdeen*. The considerable saving in coal made possible much larger ships,[33] a lesson soon learnt by the world's navies with that of Italy the first in the field with the cruiser *Dogali*, commissioned in 1884. The introduction of the triple-expansion engine signalled the beginning of the modern trans-atlantic liner, with the *City of Paris* – the first ship ever to make the crossing in under six days – and the *City of New York* entering service in 1888; both ships were also the first in their class to incorporate twin screws.[34] The principle of triple expansion could be extended to quadruple expansion, and engines of this kind were incorporated in four German liners in the period 1897–1902. This was, however, the end of the road, for by this time the steam turbine was beginning to supplant piston-driven engines.

The idea of a turbine is ancient: its basic principle is simply that of the windmill, which is essentially an instance of applied aerodynamics. Because the pressure on any solid object caused by a draught of air varies across its surface, a part of the energy of a sufficiently powerful draught encountering a sufficiently light object will be transferred to it, so that it then moves accordingly. In applied aerodynamics the profile of the object, and the material comprised in it, are chosen so as to achieve a particular type of movement. As a source of rotatory motion, the steam turbine, as its name implies, substitutes steam under pressure for air. Simply conceived of, the whole unit has axial symmetry, with steam admitted at one end through nozzles, to be exhausted at the other end after powering the rotor: just how this result is achieved is an engineering problem only solved in the 1880s by the Hon. Charles Algernon Parsons, one of the most remarkable men in the history of both technology and business.

The intuitively obvious solution is to fix blades – similar to those of a propeller – to the rotor: simplicity, however, in this case is not the answer. The problem is the high velocity 'with which steam even at only a moderate pressure escapes from a nozzle'[35]: this in turn means an unacceptably high rate of rotation, accompanied, almost inevitably, by a considerable wastage of steam. The genius of Parsons – inherited from his father, the third Earl of Rosse, who on his Irish estate had built the world's largest astronomical telescope, widely known as the 'Leviathan of Parsonstown' – consisted of realizing that the inherently high speed of a steam turbine was well suited to supplying the power to mains generators of electricity such as had been developed by Edison and Westinghouse.[36] But an industry that through most of the nineteenth century had been focused on the development and production of engines for ships and locomotives – constantly in demand by the market – was slow to turn to the requirements of electric power generation. This was where Parsons saw his opportunity.

His solution to the design problem was complex. His first turbine, completed in 1884, was so successful that the infant electricity generating industry was immediately interested. The first to be installed commercially, in 1888, was commissioned for the Forth Banks Power Station by the Newcastle and District Electric Light Company: its successful operation led to so many new orders that Parsons set up his own C.A. Parsons & Company, with works just outside Newcastle, to fulfil them. Still in his thirties he had become the founding father of a whole new industry: in the twentieth century the generation of electricity by steam power – even in nuclear power stations – far exceeded its use in any form of transport. Coal, the pre-eminent fuel for steam locomotives and steamships in the nineteenth century, is now burnt almost exclusively for the generation of electricity: somewhat paradoxically, it no longer fuels the power behind its transport, whether by land or sea, to the generating stations. As Chapter 9 shows, the electricity itself may well be used for locomotion, by road as well as rail. The essential design of Parsons' turbines still represents the state of the art.

Parsons himself – perhaps because he was a child of his times – still went ahead with designing and producing steam turbines suitable for maritime use. For this purpose he founded a second company, the Parsons Marine Steam Turbine Company. Before starting construction of the ship which he proposed to call the *Turbinia*, he constructed a six-foot-long model to be tested out on a pond on the premises of his first company. The model ship had a three-bladed propeller driven by three strands of a twisted rubber cord – a system later adopted for toy ships and aircraft. The tests then carried out enabled Parsons to calculate the required horsepower for the *Turbinia*, a ship displacing 44.5 tons that would be 100 feet long, with a beam of 9 feet and a draught of 3 feet..

The *Turbinia*, as originally constructed in 1894, had a single turbine connected directly to a single propeller. Although this could rotate at a speed of 1,730 revolutions per minute – far beyond anything that could be achieved with a reciprocating engine – a slip of 48 per cent limited the maximum speed to 19.75 knots, which was substantially less than that of the Royal Navy's most modern destroyers. Once again Parsons found the solution in increased complexity. Three turbines, in series with decreasing pressure, were installed, one on the port side, another, starboard, with the third, operating at the lowest pressure, amidships. Each turbine drove a propeller-shaft with not one, but three propellers – making nine in all. This complex system reduced the slip to 20 per cent, allowing the *Turbinia* to achieve average speeds of 25.5 knots, and maximum speeds of 31 knots, during the first sea trials in early 1897. This would be exceeded in the summer of the same year when the *Turbinia*, taking part in the Naval Review celebrating Queen Victoria's Diamond Jubilee, moved up and down the lines of warships at record speeds of up to 34 knots. As noted by the naval historian, K.T. Rowland, 'There can be few occasions in history when a fundamental new advance in technology has been demonstrated so successfully with superb confidence before so many people.'[37]

Later in the year the Institution of Naval Architects was less enthusiastic when Parsons presented a paper on the *Turbinia*, but critically he had the Admiralty on his side. In the naval arms race with France and Germany, new technology could not be disregarded, so turbines from the Parsons Marine Steam Turbine Company were ordered for two new destroyers, HMS *Cobra* and HMS *Viper*, which were ready for sea trials in 1899. Their performance exceeded all expectations, with the *Viper* achieving an average speed of 37.11 knots on one run. Tragically, in 1901, both ships were lost after foundering on rocks in bad weather. Although this double catastrophe led the Admiralty to have second thoughts, the *King Edward*, a turbine-driven merchant ship, was ordered by a new company, Turbine Steamers Ltd, in the same year.[38] The first turbine-driven liners, the *Virginian* and the *Victorian*, entered into service in 1904, with the latter being the first to cross the Atlantic. In the same year Cunard ordered the *Carmania*, its first turbine line, and in 1907 the Royal Mail Line's *Asturias* was the last ever non-turbine passenger liner. This was also the year of the sea trials of the *Mauretania* and *Lusitania*,[39] which ushered in the great age, in terms of speed, size and comfort, of ocean travel.

By this time a major transformation was taking place in the fuel burnt by ocean-going steamships. Until the 1870s there was no real alternative to coal, which in any case was abundant in the major maritime nations. Coal, even if abundant, is an inconvenient solid fuel, which is slow and dirty to load into a ship's bunkers – an operation generally considered to be almost impossible at sea. What is more, the need for bunkering on distant shores meant that special coaling stations had to be established: if ideally these should rely on local supplies, in practice this was often impossible. There were rich veins of coal in Asia and Africa, but exploiting them would depend both on railways reaching inland and the development of new harbours. To take one example, it was only at the beginning of the twentieth century that South Africa's vast coal resources in Natal were available for bunkering ships calling in at Durban. At least for Britain, with abundant coal and a railway network developed from the earliest days to transport it, it was generally cheaper to rely on these resources to supply the Empire's coaling stations: there was often no alternative.

From the 1870s onwards the position changed completely with the development of natural oil resources: on the Caspian Sea, where these were abundant – as they still are – the first ship burning oil entered service in 1878. Three years later a British yard fitted the SS (Steamship) *Gretzia* to burn oil,[40] but it was above all the Admiralty that led the way in replacing coal by oil. Although originally opposed to oil because of the risk of fire, the Royal Navy was soon won over by two considerable advantages of oil. The first, the significantly higher calorific value of oil meant useful economies in the space required by a ship's bunkers. The second – of even greater interest, given the concern to keep ships at sea – was the simple reduction in the time required for bunkering. When, in the course of the 1900s, oil-powered ships proved to be capable of spectacularly better performance, the case for oil was made. New destroyers and torpedo boats, with speeds well above 30 knots, were particularly impressive.

The conversion to oil presented the Admiralty with a new problem: a whole new infrastructure was needed to replace the established coaling stations. Although work was started almost immediately, the destroyers built under the 1908–9 building programme still burnt coal. Only by 1911 were oil-bunkering facilities sufficient, and the new Acheron-Class destroyers laid down in that year all burnt oil; the same was true of all subsequent British warships.[41]

Where the Royal Navy led, commercial shipping followed. In both cases existing ships were converted to burn oil, as well as this becoming standard for new ships. The conversion took time, and with some of the largest liners such as the *Mauretania* and the *Olympic* the process was only completed in the early 1920s.

The 1900s also witnessed another innovation even more radical in its consequences for world shipping. In 1898 the Italian inventor and entrepreneur, Guglielmo Marconi succeeded in transmitting radio signals across the English Channel. (In addition to Marconi, two of his contemporaries, Nikola Tesla and Nathan Stufflefield, took out patents for wireless radio transmitters. Tesla is now credited with being the first person to patent radio technology, following the decision of the US Supreme Court in 1943 to overturn Marconi's patent in his favour.) Marconi, unable to interest his own government in Rome in this new means of communication, set up the Marconi Telegraph Company in England, to develop it commercially. His first success in the field of ship-to-shore communications came in 1900 when the German Norddeutscher Lloyd shipping-line, intent on capturing the Atlantic Blue Riband, commissioned a wireless link with its new liner, *Kaiser Wilhelm der Grosse*. With two shore stations on the North Sea island of Borkum, 580 telegrams were handled by the end of the year. The first British ship, the *Lake Champlain*, followed in May 1901, with a wireless link to a shore station on the Isle of Wight just above the Needles rocks. By the end of the year this was just one of eight commercial stations in Britain, next to the two already operating in Germany, together with one in Belgium and two in the US.

On 12 December 1901, by transmitting a radio signal from Poldhu in Cornwall to Signal Hill in St John's, Newfoundland, Marconi conclusively demonstrated the potential of wireless telegraphy across large distances.

In 1905 radio-telegraphy enabled reports to be sent in real time of the naval battle between Russia and Japan in the Tsushima Straits separating Japan from Korea, the final decisive battle of the Russo-Japanese War (1904–5),[42] which the Russians had already lost on land when Port Arthur, their stronghold on the coast of China, fell to the Japanese at the beginning of the year. The Russian Baltic Fleet, headed by four modern battleships and commanded by Admiral Rozhdestvensky, had sailed some 8,500 miles from its home ports in a final attempt to turn the tide. This proved to be one of the most humiliating and disastrous naval expeditions in history. After barely a week Russian ships, on 21 October

1904, had fired on British fishing boats on the Dogger Bank in the North Sea, absurdly mistaking them for Japanese torpedo boats. In this action one such boat was sunk and the Russian cruiser *Aurora* was damaged as a result of what is now known as 'friendly fire'. Although the incident, which many in Britain saw as a *casus belli*, was resolved diplomatically, the Russians could not afford to lose the goodwill of a country that was, in any case, inclined to support Japan. The reason was the British concern at all the new railways being built by Russia to its Pacific coast and other remote parts of its empire in Asia: these, as related in Chapter 11, were seen as a threat to British interests in the Far East and India. After the Dogger Bank incident the British closed all their coaling stations to the Russian fleet, not caring that this might mean that it would never reach its destination in the Far East. The Russian ships were left to adopt the desperate means of bunkering at sea – from German colliers – an operation that the British Admiralty would never have contemplated. With such help the Russian fleet had almost reached its destination when it encountered the Japanese led by Admiral Heihachiro Togo in the Tsushima Straits on 27 May 1905. Although the two opposing fleets were evenly matched, Togo outmanoeuvred Rozhdestvensky and the Russians were decisively defeated, losing almost the entire Baltic Fleet, including all its eight battleships, to say nothing of 4,830 sailors killed and 8,862 taken prisoner.[43] The Battle of Tsushima, the first major sea battle in almost eighty years, was also the first ever between steamships. At the same time it was the last such battle to be fought without either side enjoying air or submarine support; with the help of wireless telegraphy the world knew of it almost the same day.

Wireless telegraphy as an adjunct to commercial steamship operations came into its own in 1909, as a result of a collision between two liners, the *Florida* and the *Republic*, off the cost of Massachusetts. On the *Republic*, which was seriously damaged and sinking, the wireless telegraphist, Jack Binns, became the hero of the hour by calling other nearby ships to come and help: this was also a triumph for Marconi, who later in the year was awarded the Nobel Prize for physics. Then, in 1910, the usefulness of his invention was highlighted by a new cause célèbre.

In the early 1900s, Harvey Hawley Crippen – an American and a self-styled medical practitioner – started to practise in London in a house shared with his domineering wife, Belle Elmore, a singer whose qualifications for opera – her chosen vocation – were as questionable as those of her husband in medicine. The household of this ill-matched couple – who made ends meet only by taking in lodgers – was in fact home to a *menage-à-trois*, the third being Crippen's lover, Ethel le Neve. Early in 1910 Belle Elmore disappeared from the scene, with her husband reporting that she had returned to the US to die – and be cremated – shortly after. Ethel le Neve lost little time in stepping into her shoes – quite literally, for she had no compunction about wearing the deceased wife's clothes and jewellery. Scotland Yard, informed of these unusual circumstances, appointed Chief Inspector Walter Dew to look into them. After having had the house searched, with nothing incriminating being found, Dew interviewed Crippen, and believed his story.

Crippen, not realizing this, panicked, and with his lover disguised as a boy, fled to Antwerp, where, under an assumed name, he booked passages on the SS *Montrose* bound for Montreal. Dew, hearing this news, had the house searched a second time, to discover the partial remains of a human body in the basement: an autopsy by Britain's best-known forensic pathologist, Bernard Spilsbury, then revealed traces of hyoscine, a drug prescribed for Belle Elmore – no doubt by her husband – which can be fatal if taken in sufficient quantities.

The news broke immediately and helped by the latest state of the art in wireless technology, Henry Kendall, captain of the *Montrose*, was among those who received it. Mingling socially with the first-class passengers – after the manner of liner captains – Kendall noticed two who aroused his suspicions. A wireless telegraph was immediately sent back to Britain: 'Have strong suspicions that Crippen London cellar murderer and accomplice are among saloon passengers. Mustache taken off growing beard. Accomplice dressed as boy manner and build undoubtedly a girl.' Dew caught a faster boat, the SS *Laurentic*, which enabled him to arrive in Quebec ahead of the Montrose. On 31 July Dew, disguised as a pilot, and supported by the Royal Canadian Mounted Police, boarded the *Montrose* and arrested both Crippen and his lover. This was the climax to a drama which the world – but not Crippen – had been able to follow from the wireless telegrams sent from both the *Montrose* and the *Laurentic* as they sailed across the Atlantic. Crippen's trial in November 1910 is remembered not only for the spectacular manner of his arrest, but also for the grisly details of how he had disposed of his wife's body: he was found guilty and hanged the same month. Ethel le Neve, tried separately, was acquitted.

In the years leading up to the First World War the greatest drama of all was the loss of the RMS *Titanic* as a result of colliding with an iceberg in the North Atlantic late in the evening of Sunday, 14 April 1912. In the years since then the scale of this disaster has, if anything, been exceeded by that of the attention that has been paid to it. The number of books describing it – the first of which appeared only two months later[44] – could well exceed that of those relating to any other historical event of such short duration. In 1997 James

Cameron's *Titanic* – the most expensive film ever made – generated such interest that within a year some seventy new books,[45] relating in one way or another to this great ship, were published. The rest of the media – radio, television and the web – kept level, as Google or any other search engine will confirm.

Although the sheer scale of the disaster goes a long way in explaining this unprecedented level of interest, it was also the result of its unlimited capacity to generate human interest stories – some of which were actually true. The wealth and distinction of the mainly American first-class passengers, and the way in which the allocation of places in the lifeboats led to many cases of only one of a couple – generally the wife on the principle of 'women and children first' – surviving, was grist to the mill for the popular press, particularly in New York, where the surviving passengers were brought after being rescued by the SS *Carpathia*. Another human interest angle came from the dedication of the *Titanic*'s two Marconi operators – as they were then known – Jack Phillips and Harold Bride, who stayed at their posts, sending out distress calls, until almost the last moment. As their ship went down they managed to jump onto an ice floe, from which Bride, though not Phillips, was finally rescued alive. Unfortunately two ships, within 20 miles of the *Titanic*, did not receive the distress signals: on one of these, the SS *Californian*, the sole Marconi operator had already gone to bed; the other, never identified, was a small steamer without any wireless apparatus.[46] On the *Carpathia*, by good fortune, the operator was still at his post, and it was this ship, some 58 miles from the *Titanic*, that in the hours after dawn on 15 April rescued all the surviving passengers. The distress signals also reached other ships that hastened to the scene of the disaster, only to find that the *Carpathia* had left no one behind to be rescued.

On the *Carpathia*, the Marconi operator, Harold Cottam, together with Harold Bride – whose feet had been frozen on the ice-floe – telegraphed the names of the survivors and the first brief reports of the catastrophe. By the time the *Carpathia* was back in New York with the surviving passengers, these reports had already proved sufficient for lurid and largely inaccurate reports in New York's tabloid press.

Notwithstanding the *Titanic* disaster's capacity to provide material for the media, in all their changing forms during the twentieth century, it is still worth noting that if the ship – as everyone expected – had successfully completed its maiden voyage, the event would hardly have been a nine-days' wonder. If at 60,000 tons the *Titanic* was for its day the largest ship afloat, it was still not as fast as the *Mauretania*, and within a generation much larger liners such as the Cunarder '*Queens*' and the French *Normandie* were built. With coal as its fuel and only one of its three propellers turbine-driven the *Titanic* was not quite 'state of the art' in all respects, although neither of these factors detracted from the drama.

If the loss of the *Titanic* served any useful purpose, it was in the attention subsequently focused on the means to enhance safety at sea. As in any such disaster bad luck played its part. The density of field ice along the course of the *Titanic* was exceptional for the time of year, although its captain had received some warning. And if the *Californian*'s Marconi operator had stayed up only a few more minutes, he would have received the *Titanic*'s distress calls: with only 20 miles between the two ships, almost everyone on the *Titanic* could have been rescued in the period of two and a half hours it took to sink. But then it was still early days for wireless communication at sea: it was, after all, only three years earlier that the collision between the *Florida* and the *Republic* provided the first occasion ever for wireless to play a key part in a rescue at sea.

In the aftermath of the *Titanic* disaster the main focus – rightly enough – was on the regulations governing the safety of ships at sea. This after all was something that lawyers, including legislators, could get hold of. Before 1912 such regulations as there were had been enacted by the governments of maritime states. Britain, with both the largest merchant fleet and the largest navy, led the way, followed by other European maritime nations, such as France, Germany and the Netherlands. The US, on the other hand, had done little in this field, and that for two reasons: first, although almost all transatlantic shipping routes ended at New York and other East Coast ports, very few of the ships that sailed them were American. Second, given the structure of American politics in the light of the division of powers between the Federal Government and the individual states, there was a certain absence of will among legislators at both levels. All this changed as a result of the *Titanic* disaster, in which several prominent Americans lost their lives. A committee of the US Senate conducted an exhaustive inquiry, and after hearing any number of witnesses found a scapegoat in the captain of the *Californian*. Naval historians now agree that this judgement was unfair: it was founded on the reports of a number of survivors that they had seen, some distance ahead of the *Titanic*, a light from another ship – which could only have been the *Californian*. If this was so, then the distress of the *Titanic* must have been visible from the *Californian*. It is now accepted that this was impossible, leaving unanswered the question as to where the light – if such it was – seen from the *Titanic* came from. It may seem odd today that Captain Smith of the *Titanic* – who went down with his ship – was completely exonerated, but no inquiry ever revealed any significant failing on his part.

With hindsight it is clear that the most serious shortcoming of the *Titanic* was to be found in its lifeboats, which had space for only 1,178 passengers – less than half the number on board the ship. Even so, the legal requirements were complied with: the last relevant legislation, the Merchant Shipping Act of 1894, had been enacted eighteen years before. In the meantime, ships – which like the *Titanic* were the pride of the lines that operated them – were built on an unprecedented scale, to travel constantly at unprecedentedly high speeds. It is easy to say now that the *Titanic*'s collision with an iceberg was an accident waiting to happen, but in the preceding ten years only nine passengers had been lost on British ships.[47] Indeed Britain had long led the way when it came to safety at sea: the mandatory Plimsoll line, that prevents the overloading of merchant ships, dated back to the Merchant Shipping Act of 1876, which also contained many other provisions relating both to safety at sea and the working conditions of ships' crews. Many such safety provisions later became mandatory internationally as a result of the International Conventions for the Safety of Life at Sea, the first of which was held in 1914 as a direct result of the *Titanic* disaster. In the context of the present history, what is important to realize is how little regulation there was before 1914: the principle of Mare Liberum, or 'freedom of the seas', propounded by the Dutch jurist, Hugo de Groot, in the seventeenth century, was recognized internationally throughout the nineteenth century era of ocean-going steamships. The seas beyond the 3-mile limit of territorial waters, agreed in 1822 by all the leading European maritime states, were in effect a legal no man's land.

The period (1865–1914) between the end of the American Civil War and the beginning of the First World War, was the great age of ocean steamships. Their passengers were not so much the high-profile wealthy and famous who travelled first class, but the millions of immigrants who travelled 'steerage' to the New World, with North America much the most favoured destination. Ellis Island, in New York harbour, where – from 1892 until 1954 – admission of new immigrants was controlled by the US Immigration Service, was the gateway to America for some 12 million people, the great majority of them arriving before the First World War. The record year of 1907, with 1,004,756 arrivals – of which 11,747 were counted on a single day, 17 April – gives an idea of the volume of shipping required. Almost every maritime nation had its flag-carriers: among many others the Cunard White Star Line, the Compagnie Générale Transatlantique and the Hamburg-Amerika line were iconic for Britain, France and Germany respectively. Beyond the Atlantic the colonial empires of the European states were each served by their own lines: the British P&O, serving India and the Far East was probably the best known, but the British travelled to South Africa by Union Castle, the French to Indochina by Messageries Maritimes and the Dutch to the East Indies by Rotterdamsche Lloyd.

The passenger liners, however, were only a part of the international shipping scene. On any measure except speed, cargo-carriers – which with the disappearance of the last of the clippers at the end of the nineteenth century were almost all steam powered – were far ahead of passenger ships. The range of different types, owners, scale of operations and land of registration, is so considerable that even a useful synoptic view is beyond the scope of this chapter. The key factor – valid for almost all shipping in the age of steam – is that the state-of-the-art technology can adapt to almost any demand: this in turn is determined by the political, economic and demographic considerations characteristic of any given stage in history. In shipping during the great age of steam that ended with the First World War, what counted above all was the absence of any rival form of transport at a time of unprecedented demand for long-distance passenger transport and for imports – much helped from the 1880s onwards by refrigerated transport by sea – from the remotest corners of the world.

Finally, the naval armaments race leading up to the First World War led to both Britain and Germany producing warships of unprecedented size and power, incorporating the latest state-of-the-art technology. In general, where Britain led, Germany followed, often with new construction proceeding at a much faster pace. These two countries were far from being alone: the US, France, Italy and, most particularly Japan, were also in the race. At least for the British, the culmination of their naval construction programme was the *Dreadnought*, which, after first being laid down in October 1905, was launched in February 1906 and ready for sea trials in October, which were completed by the end of the year.[48] The *Dreadnought* was the first large warship to be turbine driven; the range and size of its guns was beyond anything previously achieved, and all this in a ship that could steam at 21 knots. It was built in a year and a day – a record calculated to impress foreign powers: it certainly did so, at the same time encouraging them, notably Germany, to speed up their own programmes. By 1908 Germany was laying down four keels for every two laid down by Britain.[49] Even so, when it came to the state of the art, Britain generally stayed ahead, so that the cruiser HMS *Bristol*, completed in 1912, was the first warship to run its turbines on super-heated steam. Even so, in battleship design the US Nevada-Class, introduced in 1906 with the USS *South Carolina* and the USS *Michigan*, were world leaders,[50] with the last ship, the USS *Maryland* only completed in 1917. Ships of this class became known worldwide in 1941, when four of them – including the Pacific fleet's flagship, the USS *Arizona* – were the main target of the Japanese carrier-based aircraft which bombed Pearl Harbor: by this time they were long past their sell-by date, and luckily the US aircraft carriers, which were in the forefront of the naval war in the Pacific, were at sea. The great German battleships built in the years up to 1914 did not survive that long: after the surrender in 1918, the German fleet – or what was left of it after four years of war – was scuttled early in 1919, in Scapa Flow in the British Orkney Islands.

13

THE ECLIPSE OF STEAM TRANSPORT

The generation that ended with the beginning of the First World War in 1914 witnessed the maturity of transport systems – both on land and sea – powered by steam. While a traveller going back in time to the 1890s would feel quite at home in the trains and steam-ships of that time – enjoying as well a standard of personal service made possible only by the low cost of labour then prevailing – he would also marvel both at how comprehensive such transport was and how far advanced it was beyond any possible alternative.

Some time around 1890, Sherlock Holmes, accompanied by Sir Henry Baskerville, Dr James Mortimer and Dr John Watson, left London for Dartmoor on a murder investigation, travelling first class by train on Brunel's Great Western Railway – not yet converted to the standard 4-foot 8½-inch gauge[1] – to end up at a small country station that Arthur Conan Doyle's *The Hound of the Baskervilles* fails to identify.

There a 'wagonette with a pair of cobs' was waiting to bring the party to Baskerville Hall, with the first part of the journey along 'deep lanes worn by centuries of wheels'.[2] The four men would have accepted without questioning it the contrast between the essential modernity of travel by train and the antiquity of that, by a horse-drawn vehicle, which completed their journey. The speed and comfort of the former would have been compensated for by the idyllic and unspoilt countryside enjoyed from the latter.

Although today the train journey beyond Exeter would be impossible – with the track torn up and the trains replaced by buses along the A 30 main road – passengers on the express trains still running between Paddington Station in London and Exeter can still note the engineering works of Brunel – such as the bridge over the Thames at Maidenhead and the Box Tunnel in Wiltshire – on the line that he planned and constructed in the late 1830s. The experience does not differ significantly from what Sherlock Holmes's party enjoyed more than a century ago. The British railway system, already in its second half-century, was essentially mature – and considerably more extensive that it would be a century later.

Continuing in the world of detective fiction for another thirty-odd years, it is interesting to note how, in the early 1920s, guests invited to join the Duke of Denver at his shooting box on the Yorkshire moors completed their journey by dog cart – a vehicle actually drawn by a horse – after they alighted from their train from London at Northallerton Station.[3] Dorothy Sayers, whose descriptions of the Yorkshire moors in *Clouds of Witness* are almost as evocative as those of Dartmoor by Conan Doyle, does, however, portray a quite different world. The Duke of Denver, in contrast to his guests, arrived at his shooting box in his own chauffeur-driven car. What is more, his younger brother, Lord Peter Wimsey, anxious to lose no time in investigating a murder on the premises of the shooting box – for which the Duke is the prime suspect – travels by air from Paris to London. Then, at the end of the book, with key documentary evidence in his hands, he does the same from New York to an English airport – which in 1922 would have been next to impossible.[4]

From the passengers' point of view little changed in the first half of the twentieth century. When, in 1948, I travelled to Darlington on a free travel warrant at the beginning of my National Service in the British Army, the train followed the same route as that taken by the Duke of Denver's guests: with the advantage of one of the steam locomotives designed by Sir Nigel Gresley (1876–1941) in the 1930s it may have cut a few minutes off the time, but otherwise this journey on the North-Eastern main line would have been much the same as it was in Sherlock Holmes's day. The same is true of my journey to Falmouth – travelling first class along the same line, as far as Exeter, as Holmes and his party some sixty years earlier – at the end of my National Service in 1949. To all appearances there was no end in sight to the age of steam. Off the permanent way the changes were much more palpable. At the end of the 1940s there was hardly a dog cart to be seen in the North Riding or a wagonette in the West Country: the vehicles that waited for passengers at the end of their journeys were powered by the internal combustion engine.

Engine drivers, particularly on the main lines out of London, would, however, have noted some significant changes. The colour-light signalling between York and Darlington represented the most advanced state of the art as it was in the late 1930s, which, in the signal box of Northallerton – opened on 3 September 1939, the first day of the Second World War – reached its apotheosis. There, on a schematic illuminated diagram of the lines controlled from it, red lights indicated the position of every train and white lights the routes set up for trains by the signalmen. The scenario foreshadowed a modern computer terminal, with the signalmen sitting at a console where columns of switches operate the route-setting system – with both points and signals powered by electricity. There was, however, little in the technology that could not be traced back to the 1870s, and even its application to railway signalling was under way before the end of the nineteenth century. Apart from the signals there was no way of communicating with the engine drivers: telephones in the driver's cab were still some time in the future. Not only did most of the trains still run on steam, but for much of their journey they would still have been controlled by mechanical signalling systems installed in the nineteenth century – with all the characteristics described in Chapters 7 and 9.

For all its capacity to survive, the writing on the wall, suggesting the eventual decline of steam locomotion, was already to be read in Sherlock Holmes's day. Both the train in which he and his party travelled, and the station from which it departed, were lit by electric light supplied[5] – directly or indirectly – from generators based on those developed during the 1870s by Siemens and Wheatstone in Britain and Westinghouse and Edison in the US. If the members of the party had grown up in an age when this would not have been possible, by the late 1880s they must have been able to see its future potential. What is more, the potential of electricity as a source of power for transport also became apparent during the 1880s, by which time there was little doubt about the need for an alternative to steam.

London's Metropolitan Railway, which opened on 10 January 1863 as the world's first urban underground railway, worked with steam locomotives. In almost every respect the operating system was taken over, with little modification, from that of existing passenger lines. Although some of these such as the North London line carried mainly local traffic, the Metropolitan Railway, being designed exclusively for this purpose in the inner city, broke new ground. Its popularity was immediate, with an estimated 25,000 passengers on the opening day. Operating a line with only six stations, and a total journey time of just over half an hour between Farringdon and Paddington, the Metropolitan could only meet the demands of its passengers by borrowing both locomotives and rolling stock from the Great Western Railway at its Paddington terminus. On the next day, 11 January 1863, a newspaper reported that 'not only were the passengers enveloped in steam, but it is extremely doubtful if they were not subjected to the unpleasantness of smoke also.'[6] Although such trials were first blamed on the Great Western locomotives, the Metropolitan's own locomotives were equally at fault, but for many years it was to be a matter of 'what cannot be cured must be endured'. This was only possible because much of the line was in the open air – as can be confirmed by travelling the Circle Line today.

The success of the Metropolitan Railway soon led to an extension of the system in a way already provided for in the Acts of Parliament that had authorized it as far back as 1853 and 1854. The most important part of the extended system was an inner circle of twenty-seven stations, including the original six of the Metropolitan: this, known since 1949 as the Circle Line, was opened in a number of stages, with completion – long delayed by financial problems and conflicts between different companies – on 6 October 1884. By this time it had become clear not only that the disadvantages of steam made impossible the construction of lines at any depth – such as would be necessary for linking stations on both sides of the Thames – but also that electric-powered trains would solve this problem. Following the opening of the first electric tramway in Lichterfelde, just outside Berlin, in 1881,[7] the successful construction and operation of tram and trolley-bus systems in a number of different places,[8] demonstrated the potential of electric traction for trains. Here Britain led the way with the opening, in 1891, of the Waterloo and City Railway linking Waterloo Station, on the south bank of the Thames, to the Bank of England, on the north bank.

Although from the 1880s onwards the use of electric power for transport spread worldwide very rapidly, the costs involved were much too high for an across-the-board conversion of existing railways. To begin with, the fact that the actual power had to be generated outside the locomotive meant that transmission lines had to be constructed along the entire length of track; these took two forms: a third rail, which the locomotive tapped with a metal shoe, or an overhead wire tapped by a pantograph attached to the roof of the locomotive. From the earliest days both forms were used by electric trams, although for reasons of safety the third rail then had to be laid at the bottom of a cleft in the road with the power being tapped by a plough. By the 1890s the practical advantages for urban transport were such that in some cities it represented the first ever application of electric power: in Kyōto, for instance, the Sosui – a canal constructed during the 1880s to supply fresh water from Lake Biwa to the city over a distance of some twenty kilometres – was adapted to feed hydroelectric generators used exclusively for the new tramway system. Significantly, high-pressure steam – standard in British generating stations for driving turbines designed by C.A. Parsons – was not involved in Kyōto at any stage; the same was true, at much the same time, when Switzerland adopted hydro-electric power for its railways. Given the problems of smoke-abatement in the long Alpine tunnels, the need in this case was particularly acute. On the other hand the companies operating London's Inner Circle only completed the conversion to electric traction in 1905.

The final stage in the revolution made possible by electric power was its use for driving the train itself. In principle this was simple enough since the standard generator, connected to an electricity supply, becomes a motor – as Faraday had demonstrated experimentally as early as 1821.[9] The advantages of electric power were first exploited for urban transport – both trains and trams – as related earlier in this chapter.

In 1885 Karl Benz – the illegitimate son of an engine driver who died in a railway accident when Karl was two years old – constructed a *motorwagen*, the first purpose-designed vehicle to be powered by an internal combustion (IC) engine. Although the four-stroke engine used by Benz was based upon an earlier design by Nikolaus Otto, its use to power a vehicle was original, so that Benz was able to patent it a year later, in 1886. At a time when, as shown above, the potential of electric traction was beginning to be realized, few could have foreseen that the IC engine would lead not only to a far greater revolution in transport but to the growth of an industry – oil refining – that, in the twentieth century, would transform the world economy. The IC engine, as used by Benz, was not the only engine to be powered by refined petroleum: during the 1890s another German inventor, Rudolf Diesel, invented and developed the compression-ignition (CI) engine, which was fuelled by petroleum subject to two further stages of refinement, and is now known simply as diesel oil. This – as is gasoline, the fuel used by Benz in his IC engine – is a type of hydrocarbon, a name reflecting the chemical structure, based on hydrogen and carbon, of the relevant molecules.

Both gasoline and diesel oil operate by exploding within a confined space, as part of chemically distinctive vapours produced by mixing them with air. In both cases the explosion – after taking place within a cylinder – drives a piston coupled to a rotating crankshaft: the essential difference between the two systems is that with IC the explosion is produced by a spark, whereas with CI it occurs spontaneously at a certain pressure level within the cylinder – an example of the application of the gas-laws of physics. The diesel vapour explodes at a critical temperature created by pressure within the cylinder. Although the essential mechanical principle is that of the standard reciprocating steam engine, the critical operating parameters are quite different. Because the dimensions of the cylinder are far below those of a steam engine of comparable power, the size of the engine is itself much smaller in relation to the work demanded from it. This advantage is enhanced by the exceptionally high calorific value of fuel. The difference in scale is immediately apparent from comparing George Stephenson's 'Rocket' with Benz's *motorwagen*.

If in its early days the automobile developed as an adjunct to rail transport – a process characterized by taxis replacing hansom cabs on city streets – it was clear by the time of the First World War that it would be a formidable competitor. The range of the basic IC or CI engine in providing power for a vehicle was almost beyond belief: while the former was best suited for small-scale use, such as powering motorcycles, and the latter for large-scale use, such as buses and lorries, there was a considerable overlap between the two. This was the realm of the private car which, in the period between the two world wars – with the US well in the lead – became affordable for millions of households. If at this stage cars ran on petrol rather than diesel oil, it was mainly because the IC engine was cheaper and, for technical reasons, much more user-friendly – in the second half of the twentieth century new technology substantially reduced the relative disadvantage of the CI engine for use in automobiles.

In the first half of the twentieth century, the main practical limitation to the usefulness of motor transport – whether public or private – was to be found in the poor state of national road networks. Nazi Germany, by constructing the autobahn network during the late 1930s, led the way in curing this defect: after the Second World War its example was soon followed worldwide. By the 1960s it was clear that – as a result of the growth in traffic volume accompanying this process – the domain in which rail could compete with road was very much smaller than that defined by the existing rail networks, almost all of which had been built in the nineteenth century. By this time the steam engine had been replaced, according to circumstances, by diesel or electric, just as thousands of miles of uneconomical track had been abandoned worldwide.

At sea the days of steam were also numbered. With oil becoming the preferred fuel for marine steam engines at the beginning of the twentieth century, the supply infrastructure was ready for conversion from steam power to diesel. Although this first occurred with relatively small river and harbour boats, such as ferries, the *Fram*, which brought the explorer Raul Amundsen to the Arctic ice cap in 1911, was diesel powered, as was also the Danish ocean-going *Selandia*, launched a year later. The first substantial liner powered by diesel was the Swedish-American *Gripsholm*, which, after being launched in 1926, continued in service with the original engines for forty years.

Conversion to diesel was, however, a process that only took off in the second half of the twentieth century, when, as with rail transport, the advantages of diesel engines proved to be decisive. The reason behind this considerable time lag was that until the 1950s state-of-the-art marine engines could only run on high-grade diesel oil, such as was used for surface transport, whether by rail or road. This oil could not compete on costs with the low-grade 'residual' oil burnt by steamships. The introduction of CI marine engines designed to run on residual oil in the 1950s destroyed any cost advantage that steam might have had – so much so that on many ships, such as the liner, *Queen Elizabeth II*, the steam turbines were replaced by diesel engines in 1986–7. Long before this time, however, transport by sea – particularly of passengers – could no longer hold up against competing systems, which in this case meant the international airlines.

In 1903, only eighteen years after Karl Benz's first automobile, the usefulness of the IC engine proved itself, quite literally, in an entirely new dimension. On 17 December, at Kitty Hawk, North Carolina, an aeroplane designed and built by the Wright brothers, Orville and Wilbur, flew – with Orville as its pilot – a distance of just over 100 feet in 12 seconds. Because this is not a book about the air age, little more need be said about how aircraft developed to become an unrivalled means of transport for both passengers and freight. This is a history to be related in another volume. Here all that need be said is that the adaptation of steam power to aircraft, of whatever kind, was never a serious proposition. Without Karl Benz and his IC engine, the Wright brothers would have been nowhere, as they knew from the start.

If the age of steam related first to the underlying technology, as it was developed first in the eighteenth century by such inventors as Newcomen and Watt, and then in its application, during the nineteenth century to transport, by, above all, the great locomotive engineers such as Trevithick and the Stephensons, father and son, its essential economic base was the abundance of accessible reserves of coal and iron in the right places. In the case of coal, however, there were significant alternative fuels, such as – in the early days – wood in North America and Russia, and later, from the second half of the nineteenth century onwards, oil. The abundance of oil, accompanied by the development of the technology needed to refine it to produce hydrocarbon-based fuels, was one of the two dominant factors accounting for the end of the age of steam. The other was the mains supply of electricity, but only when it comes to the direct application of steam power to transport. Parsons' steam turbine is still, as it was from its early days in the 1890s, the principal engine of electric power, of which – as shown at the beginning of this chapter – transport is a major consumer. Burning hydrocarbon fuels, whether in the form of coal, oil or natural gas, accounts for much the greater part of the steam required for generating electricity: because nuclear power stations also operate by producing steam for turbo-generators, we still live in the steam age when it comes to the production of electricity, and if the future – as may well prove to be the case – lies with nuclear power, then the steam age could well outlast our time. Except, however, in special cases such as ice-breakers and submarines, this new age of steam will only be indirectly related to transport. Even then, the force of water and wind might just prove sufficient for satisfying the world's need for electricity: the fact that today's state-of-the-art technology cannot reach that far proves little. If development during the twenty-first century continues at the pace of the nineteenth or the twentieth, then who knows what may be achieved before it ends? On the other hand, it is difficult to see what could replace transport systems, whether on land or sea or in the air, dependent upon hydrocarbon fuels. These, as much as steam in the nineteenth century, define the age we now live in.

NOTES

Chapter 1

1 See Simone Martini's magnificent portrait of Guidoriccio da Fogliano (1328, Palazzo Pubblico, Siena).

2 De Rosa, L., 'Communicazioni terrestri e maritime e depressione economica: il caso del regno di Napoli (secoli XIV-XVIII)' in (ed. A.V. Marx), *Trasporti e sviluppo economico secoli XIII-XVIII*, Firenze, Felice le Monnier, 1986, pp. 3–21, p. 4.

3 This possibility is based upon an illustration in the Grimani Breviary (1505–15, Biblioteca Nazionale Marciana, Venice), for the month of September, of a four-wheeled wagon loaded with grapes and drawn by two oxen: the Flemish artist's name is unknown. The illustration for February includes two wheels, with spokes, in a shed, with a typically Dutch windmill in the background. This suggests that the wagon may be north European, with not necessarily any counterpart in Italy. The earliest pictures I have been able to find of horse and cart from Northern Europe date from the early seventeenth century: see Joos de Momper's *Winter Landscape* (1620, private collection), Rembrandt's *Farm with haystack* (1634, City

Art Museum, Copenhagen), Louis LeNain's *La Charette* (1641, Louvre, Paris) and particularly Salomon van Ruysdael's *Carriage on a road in the downs* (1631, Szépvüményszeti Múzeum, Budapest) which shows passengers being carried.

4 While Kellenbenz, H., 'Das Verkehrswesen zwischen den Deutschen Nord- und Ostseehäfen und dem Mittelmeer im 16. und in der ersten Hälfte des 17. Jahrhunderts', in (ed. A.V. Marx), op. cit., pp. 99–121, p.109 ascribes this development to the sixteenth century, in the same book as contains his article there is a photograph (following p. 27) of a wheel with spokes excavated at Gdańsk and ascribed to the twelfth century. The bas-reliefs of Borobudur, a massive Buddhist monument in Java, dating from 850 to 950, include one showing a horse-drawn passenger-carrying chariot with spoked wheels. This, however, was almost certainly unknown to Europeans until a very late stage.

5 Thomas, A.H., Hallamshire Court Rolls of the fifteenth century, pt. 4, *Trans. Hunter Arch. Soc. 11: 4* (1924), pp. 341–59 at p. 357.

6 For the remarkable antiquity of this transport strategy note the words ascribed by the bible to King Hiram of Tyre when he agreed to supply King Solomon with wood for building the temple in Jerusalem: '"My servants shall bring it down to the sea from Lebanon; and I will make it into rafts to go by sea to the place you direct ... and you shall receive it"... So Hiram supplied Solomon with all the timber of cedar and cypress that he desired': (1 Kings v: 9–10). Nothing is told about the means of land transport: Jerusalem is well inland and some 750 metres above sea level.

7 Hey, D., *Packmen, Carriers and Packhorse Roads: Trade and Communications in North Derbyshire and South Yorkshire*, Leicester University Press, 1980, p. 151.

8 Ibid., p. 205ff.

9 Trevelyan, G.M., *English Social History*, Longmans, Green & Co., 1946, pp. 249–50.

10 *The History of England*, Penguin English Library, 1979, p. 386.

11 Crump, T., *Asia-Pacific: A Modern History of Empire and War*, Continuum, 2007, p. 132.

12 Dubois, H., 'Techniques et coûts des transports terrestres en France aux XIVe et XVe siècles', in A.V. Marx (ed.), op. cit., pp. 279–291, p. 281.

13 1561, Royal Museum of Fine Arts, Brussels.

14 See Pickl, O., 'Der innereuropäische Schlachtviehhandel vom 15. bis zum 17. Jahrhundert. Routen, Umfang und Organisation', in A.V. Marx (ed.), op. cit., pp. 123–46.
15 Hey, op. cit., p. 170.
16 Vance, J.E., *Capturing the Horizon: The Historical Geography of Transportation*, Harper & Rowe, 1986, p. 384.
17 Crofts, J. *Packhorse, Wagon and Post: Land Carriage and Communications under the Tudors and Stuarts*, Routledge and Kegan Paul, 1967, Chapters X to XVII.
18 Braudel, F., *The Mediterranean and the Mediterranean World in the Age of Philip II*, Collins, 1972, Vol. I, p. 578.
19 Lane, F.C., 'Technology and productivity in seaborne transportation', in A.V. Marx (ed.) op. cit., pp. 233–44, p. 233.
20 Bagwell, P.S., *The Transport Revolution from 1770*, B.T. Batsford, 1974, p.13.
21 Deveze, M., 'Flottage en transport du bois sur les fleuves européens à l'époque moderne', in A.V. Marx (ed.), op. cit., pp.181–9. See also note 7.
22 Crump, T., *A Brief History of Science*, Constable & Robinson, 2001, p. 81.
23 'The usual way of travelling in ... most parts of the United Provinces ... is in *trek schuits*, or draw-boats, which are large covered boats, not unlike the barges of the livery companies of London, drawn by a horse at the rate of three miles an hour': Nugent, T., *The Grand Tour*, 3rd Edn., 1778, i, 48.
24 Cited in *The Oxford Companion to British History* (ed. J. Channon), Oxford University Press, 1997, p. 367.
25 Described in Blackbourne, op. cit. Pt I.
26 Vance, op. cit., p. 353.

Chapter 2

1 Hemming, J., *The Conquest of the Incas*, Macmillan, 1970, p. 101.
2 Ibid., p. 120.
3 See Chapter 1, note 15.
4 The first actual omnibus service was introduced in Nantes in 1825 by General Baudry, on whose initiative it was then extended to Paris: Vance, J.E., *Capturing the Horizon: The Historical Geography of Transportation*, Harper & Rowe, 1986, p. 355.

5 Rakmatullin, M.A., 'La situazione dei trasporti e delle vie di comuncazione nella Russia del XVIII secolo', in A.V. Marx (ed.), *Trasporti e sviluppo economico secoli XIII-XVIII*, Firenze, Felice le Monnier, 1986, pp. 227–30, p. 227.
6 Lewis, M.J.T., *Early Wooden Railways*, Routledge & Kegan Paul, 1970, p. 11.
7 The verb, *not* the metal.
8 Lewis, op. cit., following p. 72.
9 Crump, T., *Asia-Pacific*, Continuum, 2007, pp. 296–8.
10 Reproduced, Lewis, op. cit., following p. 72.
11 The 'wretched, blind, pit ponies' recalled in Ralph Hodgson's 'Bells of Heaven' (1917) reflect the common, but erroneous popular belief that a life spent underground led to blindness.
12 Ibid., pp. 310–11.
13 Ibid., p. 313.
14 Quoted, without any source given, by Ericson, S.J., *The Sound of the Whistle: Railroads and State in Meiji Japan*, Harvard University Press, 1996, p. 41
15 Quoted ibid., p. 86.
16 Ibid., p. 91.
17 Ibid., p. 150.
18 Ibid., p. 114.
19 Ibid., p. 127.
20 Ibid., p. 260.
21 Ibid., p. 177.
22 Ibid., p. 291.
23 Brace, M., 'Waterworld', in *Geographical* (1999, vol. 71., no. 4, pp. 16–22), p. 16.

Chapter 3

1 Welsh, F., *Great Southern Land: A New History of Australia*, Allen Lane, 2004, pp. 35–7.
2 Ibid., p. 48.
3 Waters, D.W., Navigational developments in the 13th to 18th centuries', in A.V. Marx (ed.), *Trasporti e sviluppo economico secoli XIII-XVIII*, Firenze, Felice le Monnier, 1986, pp. 303–9, p. 308.
4 By HM Nautical Almanac Office until May 2006, when publication was transferred to the UK Hydrographic Office: both their websites, *hmnao.rl.ac.uk* and *ukho.gov.uk*, are worth consulting.

5 Sobel, D., *Longitude*, Fourth Estate, 1995.
6 Bathurst, B., *The Lighthouse Stevensons*, Flamingo, 1999, pp. 72 and 76.
7 Lane, F.C., 'Technology and productivity in seaborne transportation', in A.V. Marx (ed.), *Trasporti e sviluppo economico secoli XIII-XVIII*, Firenze, Felice le Monnier, 1986, pp. 233–44, p. 236.
8 Braudel, F. *The Mediterranean and the Mediterranean World in the Age of Philip II*, Collins, 1972, p. 296.
9 Schurz, W.L., *The Manila Galleon*, E.P. Dutton, 1959.
10 Beaglehole, J.C., *The Life of Captain James Cook*, Stanford University Press, 1974, pp. 574–80.
11 Chaudhuri, K.N., *Trade and Civilisation in the Indian Ocean: An Economic History from the Rise of Islam to 1750*, Cambridge University Press, p. 103.
12 Private collection of Mr and Mrs Paul Mellon of Upperville Virginia. Reproduced in Walker, J., *Joseph Mallord William Turner*, Henry N. Abrahams Inc., 1976.
13 Tate Britain, London, where the visitor may compare a wide variety of Turner's magnificent seascapes.
14 *Life on the Mississippi*, Signet Classics, 1961, p. 15.
15 Jones, P., 'Italy', in M.M. Postan (ed.), *The Cambridge Economic History of Europe from the Decline of the Roman Empire*, Cambridge University Press, 1966, pp. 340–400, p. 365.
16 See also Chapter 1, note 20.
17 Brace, M., 'Waterworld', in *Geographical* (vol. 71., no. 4, pp. 16–22), 1999, p. 18.
18 Chaudhuri, op. cit., p. 139.
19 Notably Braudel op. cit.: see particularly Pt I, Section II, 'The Heart of the Mediterranean: Seas and Coasts'.

Chapter 4

1 Quoted Dickinson, H.W., *A Short History of the Steam Engine*, Frank Cass & Co, 1963, pp. 6–7.
2 Ibid., Plate II, opp. p. 37.
3 Ibid., Fig. 8, p. 48.
4 Belgium, Germany and Slovakia are the countries such as they are now: none of them were separate states in the eighteenth century.
5 Ibid., p. 50.

6 Lewis, M.J.T., *Early Wooden Railways*, Routledge & Kegan Paul, 1970, p. 205
7 Dickinson, op. cit, p. 145.
8 Ibid., p. 232.
9 Ibid., p. 57.
10 Ibid., p. 69.
11 Ibid., p. 65.
12 Quoted, ibid., p. 54.
13 Quoted, ibid., p. 89.
14 Quoted, ibid., p. 90.
15 Van Houtte, J., 'Les grandes itinéraires du commerce', in (ed. A.V. Marx), *Trasporti e sviluppo economico secoli XIII-XVIII*, Firenze, Felice le Monnier, 1986, pp. 87–97, p. 95.
16 The British National Railway Museum at York displays a vast steam engine built for this purpose.
17 Robinson, E. and Musson, A.E., *James Watt and the Steam Revolution*, Augustus M. Kelly, 1969, p. 10.
18 Dickinson, op. cit., p. 139.
19 See Chapter 2, p. 34.
20 Vance, J.E., *Capturing the Horizon: The Historical Geography of Transportation*, Harper & Rowe, 1986, p. 441.
21 *Gibbons* v. *Ogden*, 22 US 1 (1824).
22 Now in London's Science Museum. The actual builders are recorded as Hazeldine & Co, and the year of construction, 1805, so that for Trevithick it represents a relatively late design.
23 Dickinson, op. cit., p. 92.
24 Ibid., p. 95.
25 For a full description see Vance, op. cit., p. 187.
26 *Engineering*, 27 March 1868, quoted in Bagwell, P.S., *The Transport Revolution from 1770*, B.T. Batsford, 1974, p. 90.
27 The future (1820–1830) King George IV, in 1812 regent to his father, George III.
28 A city in Spain where the Duke of Wellington, in the course of the Napoleonic wars, won a crucial victory over the French Army on 22 July 1812: Blenkinsop was very much up to the mark in naming, within a month, a locomotive after this victory.
29 After the Prussian Field Marshal, whose forces, in 1815, were to play a key role, alongside those of the Duke of Wellington, in defeating Napoleon at Waterloo. See also notes 27 and 28 above.
30 Quoted Rolt, L.T.C., *George and Robert Stephenson: The Railway Revolution*, Longmans, 1960, p. 57.
31 Rolt, L.T.C., ibid., p. 37.

Chapter 5

1 Signet Classics, 1961, p. 37.
2 Vance, J.E., *Capturing the Horizon: The Historical Geography of Transportation*, Harper & Rowe, 1986, pp. 92f.
3 The middle stretch of the Grand Trunk Canal, linking the Trent and Mersey rivers systems, could only accommodate boats with a beam not greater than 2.1m, if they were to be able to pass each other; this established the narrowboat standard that is still in force: ibid, p. 92.
4 See p. 54 for the development of wagon-ways to bring the coal down to the rivers.
5 Kentucky (1792), Tennessee (1796), Ohio (1803), Illinois (1818), Missouri (1821), Arkansas (1836), Iowa (1846), Wisconsin (1848) and Minnesota (1858).
6 Two years later, in 1684, de Lasalle left France in command of a ship whose task it was to discover the mouth of the Mississippi from the sea. When, after two years, nothing was discovered, the crew mutinied and assassinated de Lasalle.
7 Hunter, L.C., *Steamboats on the Western Rivers: An Economic and Technological History*, Harvard University Press, 1949, p. 15.
8 Oliver Evans, a Western pioneer, had seen the advantages of high-pressure engines for steamboats as early as 1785 (according to his own account) but the steamboat he launched in 1803 was a failure; he had more success introducing high-pressure engines for industrial use in the West: Hunter, op. cit., p. 7.
9 Such a monopoly was declared invalid by the US Supreme Court in 1824: see Chapter 4, note 21.
10 Hunter, op. cit., p. 17.
11 Donovan, F., *River Boats of America*, Thomas E. Crowell Company, 1966, p. 52.
12 Hunter, op. cit., p. 64.
13 Ibid., p. 62.
14 Donovan, op. cit., p. 7.
15 Ibid., p. 90.
16 Twain, op. cit., p. 93.
17 The notes issued by the Banque de Citoyens of New Orleans were about the only ones worth 100 cents on the dollar: ibid., p. 93.
18 Ibid., p. 95.
19 Of these more than a thousand were on the 135-mile stretch between Baton Rouge and New Orleans: ibid., p. 92.

20 The opening of the Erie Canal in 1825 took away much of this traffic.
21 Donovan, op. cit., p. 111.
22 Author of the renowned *Birds of America*.
23 Oklahoma retained its status as Indian territory until it became the 46th state of the Union in 1907. Kansas was admitted as the 34th state in 1861 and Nebraska as the 37th in 1867.
24 Alabama, Arkansas, Florida, Georgia, Louisiana, Mississippi, North Carolina, South Carolina, Tennessee, Texas and Virginia.
25 A story told in Chapter 9.
26 Illinois, Indiana, Iowa, Kentucky, Minnesota, Missouri, Ohio, Pennsylvania and Wisconsin.
27 The name, which speaks for itself, designates the substantial lowland part of the state in which the rivers which flow into Chesapeake Bay, are tidal; in the early seventeenth century this was the first part of North America to become a permanent British settlement. From a very early stage its agricultural economy – in which tobacco was the most important crop – depended upon slave labour. The name, Virginia, was chosen to honour Queen Elizabeth I.
28 In 1863 the part of Virginia that supported the Union was admitted as the separate state of West Virginia, an event unique in US history; after the Civil War the rest of the state was only readmitted to the Union in 1870.
29 Before the construction of the Erie Canal transport of goods from Buffalo to Albany cost four times as much as transport to Montreal: Shaw, R.E., *Erie Water West: A History of the Erie Canal 1792–1854*, University of Kentucky Press, 1966, p. 101.
30 Ibid., p. 39.
31 Ibid., p. 87.
32 Ibid., p. 184.
33 Given that nearly two centuries separated Henry Hudson's discovery and exploration of the eponymous river in 1609, and the introduction of Robert Fulton's regular steamboat service in 1807, the river was for a long time home to sailing boats such as sloops and schooners, both of Dutch origin. For a time the sloops were able to withstand the competition of the steam engines of the towboat companies, but with the great increase in the size and number of cargoes, necessitating vessels of larger tonnage to transport the commodities to the New York markets with reasonable despatch and regularity, the sailing vessels of the Hudson were doomed.

34 For examples see ibid., p. 281.
35 *New York Herald*, 27 March 1846, quoted in ibid., p. 282.
36 Meaning inevitably the Great Lakes.
37 *Albany Argus*, 12 August 1845, cited ibid., p. 284.
38 Following legislative authority in 1836, the canal width was extended to 70 feet and its depth to 7 feet, while new double locks of 110 × 18 feet were constructed: ibid., p. 241. The so-called 'improved' Erie Canal was not complete along its entire length until 1862: ibid., p. x.
39 See Chapter 7, p. 134.
40 Freese, B. *Coal: A Human History*, Penguin, 2003, p. 122.
41 In a sense New Orleans already provided such a link, but its location on the Gulf of Mexico meant that ships from Europe had a much longer journey than those sailing to any of the ports on the American eastern seaboard.
42 The original charter was granted by President James Monroe in 1825, but implementing it had to wait for three years until the report required from the US Board of Engineers was received: Sanderlin, W.S., *The Great National Project: A History of the Chesapeake and Ohio Canal*, John Hopkins Press, 1946, pp. 54, 83.
43 Ibid., p. 89.
44 This link, combined with the Canadian Welland Canal linking Lake Erie to Lake Ontario, provided an alternative route for ships from the four upper Great Lakes.
45 See Chapter 6, pp. 104–110.

Chapter 6

1 Dell Publishing Co., Inc, 1960, p. 60.
2 Vance, J.E., *Capturing the Horizon: The Historical Geography of Transportation*, Harper & Rowe, 1986, p. 92f.
3 Ibid., p. 92.
4 Rowland, K.T., *Steam at Sea: A History of Steam Navigation*, Praeger Publishers, 1970, p. 13.
5 Blackbourn, D., *The Conquest of Nature: Water, Landscape and the Making of Modern Germany*, Norton, 2006, pp. 37–56.
6 Ibid., p. 94.
7 Ibid., p. 111.
8 Ibid., p. 160.

9 Ibid., pp. 156–9.
10 Ibid., p. 153.
11 Ibid., p. 154.
12 Ibid., p. 156.
13 This is the northern section of the railway, from Wadi Halfa.
14 Churchill, W.S., *My Early Life 1874–1908*, Fontana Paperbacks, 1959.
15 Ibid., p. 189.
16 Vance, op. cit., p. 454.
17 Darian, S.G., *The Ganges in Myth and History*, University Press of Hawaii, 1978, p. 93.
18 Quoted Keay, M., *Mad about the Mekong: Exploration and Empire in South-East Asia*, HarperCollins, 2005, p. 78.
19 Ibid., quoting Francis Garnier.
20 Ibid., p. 81.
21 Ibid., p. 82.
22 The main source for China is the *Cambridge Encyclopedia of China* (ed. B. Hook), Cambridge University Press, 1982, pp. 62 and 244.

Chapter 7

1 Portrayed by J.M.W. Turner in his famous picture of 1844, *Rain, Storm and Speed – The Great Western Railway*.
2 Rolt, L.T.C., *Victorian Engineering*, Allen Lane, 1970, p. 18.
3 This was a quarter of all the railway mileage in Britain at its greatest extent in the early twentieth century: Gourvish, T. R., *Railways and the British Economy 1830–1914*, Macmillan, 1980, Table IV.
4 Ibid., p. 13.
5 Quoted in Rolt, *Victorian Engineering*, p. 23.
6 Ville, S., 'Transport' in *The Cambridge Economic History of Modern Britain*, vol. 1, R. Floud and P. Johnson (eds), Cambridge University Press, 2004), pp. 295–331, p. 316.
7 Showalter, D.E., 'Railroad, the Prussian Army, and the German Way of War in the Nineteenth Century', in Nielson, K. and Otte, T.G., *Railways and International Politics: Paths of Empire, 1848–1945*, Routledge, 2006, pp. 21–44, p. 23.
8 This was granted at the Congress of Vienna in 1815, in exchange for Austria giving up the southern Netherlands: Gellner, E., *Nationalism*, Phoenix, 1998, p. 39.

9 The present station, made famous by the pictures of Claude Monet, was only constructed in the period 1886–9.
10 The New York Central, the Erie Railroad (quite separate from the canal but built with the same economic rationale), the Pennsylvania Railroad and the B&ORR.
11 The Atlantic & St Lawrence Railroad between Portland Maine (in the US) and Montreal.
12 Nielson, K. and Otte, T.G., *Railways and International Politics: Paths of Empire, 1848–1945*, Routledge, 2006, p. 10.
13 Ibid., p. 11.
14 See also Chapter 5, p. 90.
15 The story of the Andrews raid, as related in this chapter, is based on Bonds, R.S., *Stealing the General: The Great Locomotive Chase and the First Medal of Honor*, Westholme Publishing, 2007.

Chapter 8

1 Dickens, Charles, *Dombey and Son*, Penguin Classics, 1985. The construction of new railways in North London is a central theme of this novel of the 1840s.
2 Cited Rolt, L.T.C., *George and Robert Stephenson: The Railway Revolution*, Longmans, 1960, p. 63.
3 Cited ibid., p. 111.
4 Ibid. p. 154.
5 Ibid., p. 155.
6 Cited, ibid., p. 158.
7 For this, and the citation in the previous paragraph, see ibid., p. 160.
8 Ibid., p. 164.
9 Ibid., p. 175.
10 Ibid., p. 181.
11 Needless to say Stamford got its railway in the end, on the cross-route linking East Anglia to the Midlands.
12 The first line was from Mechelen to Brussels South; this, the oldest city terminal in mainland Europe, is now also the terminus for the EuroStar trains to London. Belgium has long had the world's densest railway network.
13 Crump, T., *A Brief History of Science*, Constable & Robinson, 2001, p. 88.

14 Rolt, L.T.C., *Red for Danger: A History of Railway Accidents and Railway Safety*, David & Charles, 1982, p. 41.
15 John Saxby registered the first British patent for Interlocking of Points and Signals in 1856. Then in 1860 Saxby & Farmer, of which he was co-founder, became the world's first railway-signalling manufacturer.
16 First installed on the London and Croydon Railway in 1841.
17 Crump, op. cit., pp. 155–8.
18 Ibid., p. 112.
19 Rolt, op. cit., p. 30.
20 See ibid., Chapter 2, 'Double Line Collisions of Early Days'.
21 Published by Butterworth.
22 This was the result of a long-established legal principle known as the 'Rule against Perpetuities'.
23 'Transport' in *The Cambridge Economic History of Modern Britain*, vol. 1, Floud and P. Johnson (eds), Cambridge University Press, 2004), pp. 295–331, p. 315.
24 Gourvish, op. cit., p. 13.
25 This part of the 1844 Act was replaced in 1883 by the Cheap Trains Act.
26 Gourvish, op. cit., p. 52. See also page 178 above.
27 *Priestley* v. *Fowler* (1837) 3 M. & W. 1. This decision was approved by the House of Lords in *Bartonshill Coal Co.* v. *Reid* (1858) 3 Macq. 266.
28 *Farwell* v. *Boston and Worcester Railway* (1841) 38 Am. Dec. 339.
29 Law Reform (Personal Injuries) Act, 1948.
30 Alderman, G., *The Railway Interest*, Leicester University Press, 1973, p. 177
31 Gourvish, op. cit., p. 53.
32 [1901] A.C. 246.

Chapter 9

1 Rolt, L.T.C., *Victorian Engineering*, Allen Lane, 1970, p. 183.
2 Ibid., p. 192f.
3 Although this was only introduced in 1898, its predecessor, the 4–4–0 dates back to the late 1870s: Rolt, op. cit., p. 198.
4 Quoted, with no source given, by Ericson, S.J., *The Sound of the Whistle: Railroads and State in Meiji Japan*, Harvard University Press, 1996, 78, where it is also noted that the

American Pennsylvania Railroad was, in 1878, the first to introduce lavatories.

5 Named after its inventor Eli H. Janney.

6 Busch became famous for his Budweiser beer, named after the town where it was first brewed in what is now the Czech Republic. This means that Busch's Budweis is now correctly named Budjovice.

7 Crump, T., *A Brief History of Science*, Constable & Robinson, 2001, p. 116.

8 Rolt, op. cit., p. 226f.

9 Vance, J.E., *Capturing the Horizon: The Historical Geography of Transportation*, Harper & Rowe, 1986, pp. 257–8.

10 Hobsbawm, E.J., *Industry and Empire: An Economic History of Britain since 1750*, Weidenfeld and Nicolson, 1968, p. 93.

11 Morris, J., *Pax Britannica: The Climax of an Empire*, Penguin Books, 1979, p. 369, who notes that the first goods train to cross the continent carried stores for the Pacific Squadron of the Royal Navy.

12 The full name is the Aitchison, Topeka and Santa Fe.

13 'What the Railroad Will Bring Us', *Overland Monthly* (vol. 1), 1868, p. 303.

Chapter 10

1 Morris, J., *Pax Britannica: The Climax of an Empire*, Penguin Books, 1979, p. 169.

2 This, and the following quotation, are taken from Satow, M. and Desmond, R., *Railways of the Raj*, Scolar Press, 1980, p. 13: this book is the main source of the material in the present section.

3 Ibid. p. 19.

4 Kerr, I.J. *Building the Railways of the Raj 1850–1900*, Oxford University Press, 1995, p. 1.

5 MacGeorge, G.W., *Ways and Works in India*, quoted in Satow and Desmond, op. cit., p. 21.

6 Ibid., p. 45.

7 Ibid., p. 39.

8 Ibid., p. 43.

9 Later to become King George V and Queen Mary.

10 A photograph, relating back to the royal visit, appeared in the *Illustrated London News* of 20 January 1906.

11 Quoted in Lewis, C.M., *British Railways in Argentina, 1857–1914*, Athlone, 1983, p. 17.
12 Ibid., p. 225, note 20.
13 Hutchins, J.G.B., *The American Maritime Industries and Public Policy, 1789–1914: An Economic History*, Harvard University Press, 1941, p. 341. The voyage from Europe took about ten weeks.
14 Wright, W.R., *British-owned Railways in Argentina: Their Effect on the Growth of Economic Nationalism, 1854–1948*, University of Texas Press, 1974, p. 15.
15 Lewis, op. cit., p. 7, Wright, op. cit., p. 24.
16 Trevelyan, G.M., *English Social History*, Longmans, Green & Co., 1946, p. 536.
17 Wright, op. cit., p. 45.
18 Ibid. p. 48.
19 Ibid. p. 61.
20 Quoted ibid., p. 71.
21 Quoted ibid., p. 85.
22 The origins of this familiar phrase seem to be lost in time: it is the title of a book published by Pete Arno in 1956, and also occurs in one of his *New Yorker* cartoons of the same year.
23 Wright, op. cit., p. 106, quoting the *Buenos Aires Standard*.
24 Quoted ibid. p. 103.
25 Eva, known affectionately as Evita, died in 1952, aged only 33, leaving her husband to rule without her popular support – commemorated in an opera by Andrew Lloyd-Webber. Juan Perón returned briefly to power in 1974, to be succeeded on his death by his third wife, Isabel. After a failed administration, she was deposed by a military junta in 1976, which in 1982 was finally defeated in its attempt to take over the Falkland Islands from the British.
26 Quoted Welsh, F., *Great Southern Land: A New History of Australia*, Allen Lane, 2004, p. 536, who goes on to add that 'If a "banana republic" is taken to mean an economy which relies upon the export of primary commodities to finance its trade, the description did in fact aptly describe Australia.'
27 Morris, J., *Pax Britannica: The Climax of an Empire*, Penguin Books, 1979, p. 366, note 2. The following pages survey railway construction throughout the British Empire.
28 Pakenham, T., *The Boer War*, Futura, 1982, p. 102. This book has been my main source of the history of the war.
29 Ibid., p. 422.

Chapter 11

1 Westwood, J.N., *A History of Russian Railways*, George Allen and Unwin, 1964, pp. 22–3.
2 Ibid., p. 24.
3 Ibid., p. 28–9.
4 See L. Olpiphant's letter, quoted in ibid., pp. 36–7.
5 Heywood, A.J., 'The most catastrophic question: railway development and military strategy in late imperial Russia', in K. Nielson and T.G. Otte (eds), *Railways and International Politics: Paths of Empire, 1848–1945*, Routledge, 2006, pp. 45–67, p. 45.
6 Westwood, op. cit., p. 71.
7 Ibid., p. 124.
8 Ibid., pp. 116–19.
9 Ibid., p. 114.
10 Ibid., p. 115.
11 Between 1917 and the beginning of the Pacific War in 1941 the preferred fast route to Japan was to cross both the Atlantic and Pacific by boat, and North America by train. This position was restored after the defeat of Japan in 1945, but from the 1950s onwards almost all passengers went by air. Gladys Aylward's journey to China in the mid 1930s, as portrayed by Ingrid Bergman in Mark Robson's film, *Inn of the Sixth Happiness* (1958), gives a brief glimpse of travel on the Trans-Siberian Railway.
12 Westwood, op. cit., p. 122.
13 Harries, M. and S., *Soldiers of the Sun: the Rise and Fall of the Imperial Japanese Army*, New York, Random House, 1991, p. 57.
14 *Cambridge Encyclopedia of Japan*, 1982, p. 250.
15 Kent, P.H., *Railway Enterprise in China: An Account of its Origin and Development*, London, 1907, p. 93.
16 Otte, T.G., 'The Baghdad Railway of the Far East': The Tientsin-Yangtze railway and Anglo-German relations, 1908–1911, in Nielson, K. and Otte, T.G., *Railways and International Politics: Paths of Empire, 1848–1945*, Routledge, 2006, pp. 112–36, p. 113.
17 Ibid., p. 119.
18 Preston, D., *The Boxer Rebellion*, Constable & Robinson, 2002, p. 14.
19 Ibid., p. 19.
20 Otte, op. cit., p. 114.
21 Ibid., pp. 116–17.

22 Quoted in ibid., p. 115.
23 Ibid., p. 121.
24 Preston, op. cit., pp. 84, 325.
25 Otte, op. cit., p. 123.
26 Quoted in ibid., p. 127.
27 Morison, S.E., *'Old Bruin': Commodore Matthew C. Perry 1794–1858*, Boston, Little, Brown & Co., 1967, p. 4.
28 Quoted in Ericson, op. cit., p. 4.
29 Inoue Masaru, cited in ibid., p. 25.
30 Cullen, L.M., *A History of Japan, 1582–1941: Internal and External Worlds*, Cambridge University Press, 2003, pp. 199–204.
31 Ericson, op. cit., p. 39.
32 'TÿÿM' means 'east' and 'zaï', west: just think of Essex and Wessex.
33 Ericson, op. cit., p. 54.
34 Ibid., p. 62.
35 Ibid., p. 58.
36 Ibid., p. 66.
37 Ibid., p. 68.
38 Ibid., p. 81.
39 Ibid., p. 76.
40 Ibid., p. 86.
41 Ibid., p. 51.
42 Quoted in ibid., p. 49.
43 Ericson, op. cit., pp. 40–1.
44 Ibid., pp. 81–4.
45 Ibid., p. 92.

Chapter 12

1 Rowland, K.T., *Steam at Sea: A History of Steam Navigation*, Praeger Publishers, 1970, p. 45.
2 Ibid., p. 52.
3 Ibid., p. 50.
4 Ibid., p. 55.
5 Ibid., p. 52.
6 Ibid., p. 69.
7 Rowland, op. cit., p. 64.
8 Ibid., pp. 65–6.
9 Ibid., p. 73.

10 For the trials of Hall and his condenser, see Rowland, op. cit., pp. 105–6.
11 Rowland, op. cit., p. 77.
12 Ibid., p. 75.
13 Letter to *The Times*, 24 October 1844, from F. Collier Christy of the Thames Iron Works.
14 Buchanan, A., *Brunel: The Life and Times of Isambard Kingdom Brunel*; Continuum, 2006, p. 60.
15 Ibid., p. 61.
16 Rowland, op. cit., p. 79.
17 Quotations from Buchanan, op cit., p. 117.
18 Ibid., p. 118.
19 *The Times*, 21 September 1859 (repeating the claim already made in 1857) quoted in Buchanan, op. cit., p. 132.
20 Rowland, op. cit., p. 161.
21 Rowland, op. cit., p. 145.
22 Rowland, op. cit., p. 119.
23 *The Oxford Companion to Ships and the Sea* (eds I.C.B. Dear and P. Kemp), Oxford University Press, 2006, p. 127.
24 See *Oxford Concise Science Dictionary*, 3rd ed., 1996, p. 302.
25 Rowland, op. cit., p. 129.
26 Ibid., p. 120.
27 Ibid., p. 92.
28 Hutchins, J.G.B., *The American Maritime Industries and Public Policy, 1789–1914: An Economic History*, Harvard University Press, 1941, p. 456.
29 Ibid., p. 373.
30 Rowland, op. cit., pp. 132–3.
31 Hutchins, op. cit., p. 456.
32 Ibid., p. 492.
33 Rowland, op. cit., p. 153.
34 Ibid., p. 129.
35 Rowland, op. cit., p. 169.
36 See Chapter 9, p. 193.
37 Rowland, op. cit., p. 172.
38 Hutchins, op. cit., p. 491.
39 Ibid., p. 503 and Rowland, op. cit., p. 17.
40 Ibid., p. 184.
41 Ibid., p. 185.

42 See Harries, M. and Harries, S., *Soldiers of the Sun: the Rise and Fall of the Imperial Japanese Army*, New York, Random House, 1991, Chapter 8.
43 Ibid., p. 91.
44 This is probably Lawrence Beesley's *The Loss of the S.S. Titanic by one of the Survivors*, Houghton Mifflin Co, 2000.
45 I was given this information at Falmouth's Maritime Bookshop early in 1999.
46 Ibid., p. 116.
47 Ibid., p. 169.
48 Parkes, O., *British Battleships: 'Warrior' 1860 to 'Vanguard' 1950: A History of Design, Construction and Armament*, Seeley Service & Co, 1957, Chapter 81.
49 Ibid., p. 508.
50 Ibid., p. 579.

Chapter 13

1 A walking-stick 'of the sort which is known as a "Penang Lawyer" was presented to Dr. Mortimer in 1884, some five years before he came to ask the help of Sherlock Holmes': Conan Doyle, A., *The Hound of the Baskervilles*, Harper Brothers, 1901, Chapter 1, 'Mr Sherlock Holmes'. The GWR converted to the standard gauge in 1892. It is surprising that a book first published in 1901 should recount events, even if fictional, so far in the past.
2 Quoted from ibid., Chapter 6, 'Baskerville Hall'. 'Centuries' is probably an exaggeration.
3 Dorothy Sayers also refers to the 'ostler' – a word unknown to the Microsoft Word spell check – employed by the hotel in the fictional local market town of Stapley, which can be identified with Helmsley, where the Black Swan hotel is much to be recommended.
4 Although nothing in *Clouds of Witness* enables a precise year to be given to the events it relates, they are referred to in Dorothy Sayers' *Strong Poison*, which relates specifically to 1927, as having occurred five years previously.
5 The Institution of Engineering and Technology offers a portfolio of 16 mounted photographs entitled 'Electric Lighting for Paddington Station J E H Gordon's System, 1885'.

6 'Opening of the Metropolitan Railway to the public', Sunday January 11, 1863, republished in *Guardian Unlimited*.
7 In 1880 the Russian, Fiodor A. Pirotskiy, was the first to invent and test an electric tram.
8 In 1888 Richmond, Virginia, became the first American city with electric trams or 'trolleys'.
9 Crump, op. cit., pp. 113f. The motor effect was in fact discovered first, and from observing it Faraday deduced the possibility of reversing it so as to generate electricity.

BIBLIOGRAPHY

Alderman, G., *The Railway Interest*, Leicester University Press, 1973.

Bagwell, P.S., *The Transport Revolution from 1770*, B.T. Batsford, 1974.

Bathurst, B., *The Lighthouse Stevensons*, Flamingo, 1999.

Beaglehole, J.C., *The Life of Captain James Cook*, Stanford University Press, 1974.

Beesley, L., *The Loss of the S.S. Titanic by one of the Survivors*, Houghton Mifflin Co, 2000.

Blackbourn, D., *The Conquest of Nature: Water, Landscape and the Making of Modern Germany*, Norton, 2006.

Bonds, R.S., *Stealing the General: The Great Locomotive Chase and the First Medal of Honor*, Westholme Publishing, 2007.

Brace, M., 'Waterworld', in *Geographical* (vol. 71., no. 4, pp. 16–22), 1999.

Braudel, F., *The Mediterranean and the Mediterranean World in the Age of Philip II*, Collins, 1972.

Buchanan, A., *Brunel: The Life and Times of Isambard Kingdom Brunel*, Continuum, 2006.

Cambridge Economic History of Europe, vol. VI, 'The Industrial Revolutions and After', CUP, 1966.

Cambridge Encyclopedia of China, B. Hook (ed.), Cambridge University Press, 1982.

Churchill, W.S., *My Early Life 1874–1908*, Fontana Paperbacks, 1959.

Cole, A.H. and Williamson, H.F., *The American Carpet Manufacture: A History and Analysis*, Harvard University Press, 1941.

Conan Doyle, A., *The Hound of the Baskervilles*, Harper Brothers, 1901.

Conrad, J., *The Heart of Darkness*, Dell Publishing Co., Inc, 1960.

Crofts, J. *Packhorse, Wagon and Post: Land Carriage and Communications under the Tudors and Stuarts*, Routledge and Kegan Paul, 1967.

Crump, T., *A Brief History of Science*, Constable & Robinson, 2001.

——, *Asia-Pacific*, Continuum, 2007.

Cullen, L.M., *A History of Japan, 1582–1941: Internal and External Worlds*, Cambridge University Press, 2003.

Darian, S.G., *The Ganges in Myth and History*, University Press of Hawaii, 1978.

Dickens, C., *Dombey and Son*, Penguin Classics, 1985, pp. 120–1.

Dickinson, H.W., *James Watt: Craftsman and Engineer*, August M. Kelley, 1967.

——, *A Short History of the Steam Engine*, Frank Cass & Co, 1963.

Donovan, F., *River Boats of America*, Thomas E. Crowell Company, 1966.

Ellis, C.H., *The Lore of Steam*, Hamlyn Paperbacks, 1983.

Ericson, S.J., *The Sound of the Whistle: Railroads and State in Meiji Japan*, Harvard University Press, 1996.

Falconer, J. *Brunel*, Ian Allen, 2005.

Freese, B. *Coal: A Human History*, Penguin, 2003.

Gellner, E., *Nationalism*, Phoenix, 1998.

George, H., 'What the Railroad Will Bring Us', *Overland Monthly*, vol. 1, 1868, p. 303.

Gough, R., *The History of Myddle*, Penguin, 1981.

Gourvish, T.R., *Railways and the British Economy 1830–1914*, Macmillan, 1980.

Grey, R. *A History of London*, Hutchinson, 1978.

Harries, M. and S., *Soldiers of the Sun: the Rise and Fall of the Imperial Japanese Army*, New York, Random House, 1991.

Hemming, J., *The Conquest of the Incas*, Macmillan, 1970.

Hey, D., *Packmen, Carriers and Packhorse Roads: Trade and Communications in North Derbyshire and South Yorkshire*, Leicester University Press, 1980.

Heywood, A.J., 'The most catastrophic question': railway development and military strategy in late imperial Russia', in T.G. Otte and K. Nielson (eds), *Railways and International Politics: Paths of Empire, 1848–1945*, Routledge, 2006, pp. 45–67.

Hobsbawm, E.J., *Industry and Empire: An Economic History of Britain since 1750*, Weidenfeld and Nicolson, 1968.

Hunter, L.C., *Steamboats on the Western Rivers: An Economic and Technological History*, Harvard University Press, 1949.

Hutchins, J.G.B., *The American Maritime Industries and Public Policy, 1789–1914: An Economic History*, Harvard University Press, 1941.

Jackson, G. and Williams, D.M. (eds), *Shipping Technology and Imperialism*, Scolar Press, 1996.

Jones, P., 'Italy', in *The Cambridge Economic History of Europe from the Decline of the Roman Empire*, M.M. Postan (ed.), Cambridge University Press, 1966, vol. I, pp. 340–431.

Keay, M., *Mad about the Mekong: Exploration and Empire in South-East Asia*, HarperCollins, 2005.

Kellenbenz, H., 'Das Verkehrswesen zwischen den Deutschen Nord- und Ostseehäfen und dem Mittelmeer im 16. und in der ersten Hälfte des 17. Jahrhunderts', in A.V. Marx (ed.), *Trasporti e sviluppo economico secoli XIII-XVIII*, Firenze, Felice le Monnier, 1986.

Kent, P.H., *Railway Enterprise in China: An Account of its Origin and Development*, London, 1907.

Kerr, I.J. *Building the Railways of the Raj 1850–1900*, Oxford University Press, 1995.

Lewis, C.M., *British Railways in Argentina, 1857–1914*, Athlone, 1983.

Lewis, M.J.T., *Early Wooden Railways*, Routledge & Kegan Paul, 1970.

Macaulay, Lord, *The History of England*, Penguin English Library, 1979.

Morison, S.E., *'Old Bruin': Commodore Matthew C. Perry 1794–1858*, Boston, Little, Brown & Co.,1967

Morris, J., *Heaven's Command: An Imperial Progress*, Penguin Books, 1979.

——, *Pax Britannica: The Climax of an Empire*, Penguin Books, 1979.

Nielson, K. and Otte, T.G., *Railways and International Politics: Paths of Empire, 1848–1945*, Routledge, 2006.
Oxford Concise Science Dictionary, 3rd ed., 1996.
Pakenham, T., *The Boer War*, Futura, 1982.
Parkes, O., *British Battleships: 'Warrior' 1860 to 'Vanguard' 1950: A History of Design, Construction and Armament*, Seeley Service & Co, 1957.
Rakmatullin, M.A., 'La situazione dei trasporti e delle vie di comunicazione nella Russia del XVIII secolo', in A.V. Marx (ed.), *Trasporti e sviluppo economico secoli XIII-XVIII*, Firenze, Felice le Monnier, 1986, pp. 227–30.
Robinson, E. and Musson, A.E., *James Watt and the Steam Revolution*, Augustus M. Kelly, 1969.
Rolt, L.T.C., *George and Robert Stephenson: The Railway Revolution*, Longmans, 1960.
——, *Red for Danger: A History of Railway Accidents and Railway Safety*, David & Charles, 1982.
——, *Victorian Engineering*, Allen Lane, 1970.
Rowland, K.T., *Steam at Sea: A History of Steam Navigation*, Praeger Publishers, 1970.
Sanderlin, W.S., *The Great National Project: A History of the Chesapeake and Ohio Canal*, John Hopkins Press, 1946.
Satow, M. and Desmond, R., *Railways of the Raj*, Scolar Press, 1980.
Sayers, D.L., *Clouds of Witness*, Hodder & Stoughton, 2003.
——, *Strong Poison*, Avon Books, 1967.
Schurz, W.L., *The Manila Galleon*, E.P. Dutton, 1959.
Shaw, R.E., *Erie Water West: A History of the Erie Canal 1792–1854*, University of Kentucky Press, 1966.
Showalter, D.E., 'Railroad, the Prussian army, and the German way of war in the nineteenth century', in Nielson, K. and Otte, T.G., *Railways and International Politics: Paths of Empire, 1848–1945*, Routledge, 2006, pp. 21–44.
Sobel, D., *Longitude*, Fourth Estate, 1995.
Taylor, E.G.R., *The Haven-Finding Art: A History of Navigation from Odysseus to Captain Cook*, Hollis & Carter, 1956.
The Oxford Companion to British History, J. Channon (ed.), Oxford University Press, 1997.
The Oxford Companion to Ships and the Sea, I.C.B. Dear and P. Kemp (eds), Oxford University Press, 2006.

Thomas, A.H., 'Hallamshire Court Rolls of the fifteenth century', pt. 4, *Trans. Hunter Arch. Soc. 11: 4* (1924), pp. 341–59.

Trevelyan, G.M., *English Social History*, Longmans, Green & Co., 1946.

Vance, J.E., *Capturing the Horizon: The Historical Geography of Transportation*, Harper & Rowe, 1986.

Van Houtte, J., 'Les grandes itinéraires du commerce', in A.V. Marx (ed.), *Trasporti e sviluppo economico secoli XIII-XVIII*, Firenze, Felice le Monnier, 1986, pp. 87–97.

Ville, S., 'Transport', in *The Cambridge Economic History of Modern Britain*, vol. 1, R. Floud and P. Johnson (eds), Cambridge University Press, 2004, pp. 295–331.

Walker, J., *Joseph Mallord William Turner*, Henry N. Abrahams Inc., 1976.

Wallis, M., *Canaletto Malarz Warszawy*, PIW Warsaw, 1955.

Waters, D.W., 'Navigational developments in the 13th to 18th centuries', in A.V. Marx (ed.), *Trasporti e sviluppo economico secoli XIII-XVIII*, Firenze, Felice le Monnier, 1986, pp. 303–9.

Welsh, F., *Great Southern Land: A New History of Australia*, Allen Lane, 2004.

Westwood, J.N., *A History of Russian Railways*, George Allen and Unwin, 1964.

Wright, W.R., *British-Owned Railways in Argentina: Their Effect on the Growth of Economic Nationalism, 1854–1948*, University of Texas Press, 1974.

INDEX